CULTURAL
DIVERSITY

A GEOGRAPHIC PERSPECTIVE

LEONARD G. PEACEFULL

University of Akron

Kendall Hunt
publishing company

Kendall Hunt
publishing company

www.kendallhunt.com
Send all inquiries to:
4050 Westmark Drive
Dubuque, IA 52004-1840

Copyright © 2011 by Kendall Hunt Publishing Company

ISBN 978-0-7575-9787-9

Printed in the United States of America
10 9 8 7 6 5 4 3 2

Contents

Contents

Foreword

For the last three centuries or so, the idea of Cultural Diversity has been slowly yet steadily strengthening. Initially, as a viewpoint of some academic scholars and perceptive, thoughtful political leaders, the included concepts have matured as an increasingly significant explanatory force for world problems and development: population expands while the space or size of the world does not. Yet it is only with the recent growth of world-wide communications and the increasing ease of human group contacts, that it has engendered profound significance. Mankind has begun to understand dramatically that the earth's "worlds beyond" exist, and that their human resources affect each and every one of us, both negatively and positively.

Clichés such as "globalization," "coexistence" and "global village" have become part of the speech of those writers and political leaders who recognize that world peace and security will come to fruition only if one ethnic group seeks to know and understand another. At the same time, those who do not understand the importance of cultural diversity, those who fear or are uncomfortable with the outlook, the expression and the orientation of peoples of other ethnic designations, help foster suspicion and hostility in their larger populations.

One of the most important benefits of the study of cultural diversity, is a gradual growth of better understanding of our own culture, and the limitations that it has produced in the way we live, in how we say things, the expressions we prefer to use, the values we place, often without thinking, on the values of others and perhaps even the way we dress and the kinds of food we eat.

What better way to examine cultural diversity than through a world-wide view of the specific ethnic and cultural patterns. This volume reveals and places these factors solidly as products of the processes which provide the greatest relationship between groups of people and the environment. Understanding cultural diversity does not mean that one accepts every element of other cultures, but it does offer the educational opportunity and benefit of understanding why other people act the way they do. This excellent volume offers readers the unique opportunity to view the broad sweep to world cultures from the perspectives of three outstanding geographers from three different cultures—North American, Asian, and European.

And, always remember to judge not, without careful thought and reasoned perception.

Allen G. Noble
Distinguished Professor of Geography and Planning

Preface

This text is directed to students coming to a cultural diversity course for the first time. It aims to put cultural diversity and the related conflicts in a World context. Through an awareness of the various languages, beliefs, attitudes and customs can we can begin to understand the diversity of cultures in different parts of the world, showing that between and within different areas of the world there is a great diversity of cultures.

Cultural diversity can lead to conflict between and within cultures, a concept which may be difficult to understand from a western perspective. The world today is changing. Globalization is the new jargon. Through the media of the internet and its associated messaging facilities, such as email, facebook, twitter, etc., we are closer to people on the other side of the World than we have ever been. We communicate and integrate with other cultures on a daily basis, yet as Americans we tend to look at the world and relate what we see to our own experience. In many cases our perceptions are false, which leads us to make judgments that are unfounded. To truly understand cultures we must try to put ourselves in the place of the culture and experience life as these people undergo daily trials and tribulations.

In reading this text you will observe that the cultural differences are liable to result in conflicts of some kind or another. It is hoped that you will realize that these cultural differences have a more profound history than our current perception. Consequently, it will be clear that an appreciation of the historical geography is just as significant as a grasp of the present-day cultural geography. Cultural geography is a subfield of human geography that focuses specifically on the role of human cultures, languages, beliefs, customs, artifacts, etc., and their interactions and influence on the landscape.

Why a Geographic Perspective?

Geography is an integrated discipline that brings together the physical and human dimensions of the world. For me, it is about how we see the world from a physical and a human sense. It takes the physical aspects and relates them to human activity. Geographers study real places and real people and real events that affect those places and people. It is this broad interrelationship between humans and their environment that makes geography such a fascinating subject. To many outsiders, geography is "knowing" the capital of such and such or the longest river in somewhere or the highest mountain in wherever. If you need that information, go look it up in an atlas—one of the many tools of a geographer.

Geography is a great subject for lateral thinkers because there are no boundaries. Going sideways is just as important as going forward (Gale, 1993). Concern with locations in space is a characteristic of a geographer's curiosity; specifying location accurately is one of the prime rules of the geographic game. An inaccurate description of location causes a geographer to shudder in the same way a linguist would at a mispronunciation or a musician when hearing a wrong note.

We are all geographers. We are confronted with spatial relationships almost daily. As humans we have an innate sense of spatial reality, some can even remember a route after only taking it once or twice. For example, you are probably aware of the relationships between different buildings on campus and after one or two expeditions can make your way easily between them. Likewise, we may go into a store and know the geography of that store even in a different town or location. Those of you who are familiar with Wal-Mart, Target, or even Home Depot will know that one store is laid out very much like another. This is a basic tool of retail, keep the geography of the store similar and customers will feel comfortable shopping in the store.

Consequently, geography offers a way of thinking about problems on a local, regional, or world scale. Geographers are well equipped to understand interactions among different forces affecting a place. For example, to explain cultural conflicts in various parts of the world we study the interrelationships among the distribution of resources, the differences in religious beliefs, the historical settlement patterns, the international ties of local groups and the conflicting strategies for achieving economic growth. The interrelationships among factors affecting places help us understand why humans behave as they do. No other scientific discipline takes this approach.

The Format of the Book

The 10 chapters of the text can be divided into five sections. In the first section we look at the diversity and development of cultures. This is followed by three chapters that together examine the impact humankind has on the landscape through the development of the cultural landscape, agriculture and urbanization. The third section deals with the underlying themes of ethnicity, their origins, and diffusion. Chapter six looks at languages, the many and varied language families and the diffusion of the English language; while chapter seven explores the origin and development of beliefs and includes a review of the world's major religions. The fourth section deals with both ethnic issues and gender issues. Chapter eight is a brief analysis of some of the civil wars that have recently taken place or are ongoing, followed by an overview of the dilemma of ethnic groups who have no politically independent territory. Chapter nine continues the conflict theme but examines the plight of women throughout the world. Though we have concentrated mostly on Asia, there are certainly other areas that can be studied. Finally, the last section, chapter ten, takes all of the themes of the previous chapters and relates them—North America and the United States in particular. In this section, we attempt to answer the question: What makes the American culture distinctive?

Leonard Peacefull
Akron, Ohio
June 2011

CHAPTER 1

The Diversity of Cultures

INTRODUCTION

In a world of seven billion people there exist a number of cultures, each having some variation from the next. These many disparities include customs, beliefs, ideals and attitudes. At times we tend to look at the world and relate what we see to our own experience. Unfortunately, in many cases our perceptions are false. Taking a false perspective can and has over time, led to confusion. These diversities between and sometimes within, cultures can also sometimes trigger conflict.

Examination of a world map shows the world to be divided politically into countries and territories each with their defined boundaries. In regard to the distribution of cultures this sometimes conveys a misleading image of people's cultural identity. For example, most people living in France would consider themselves to be French, while a small minority may think of themselves as Bretons, or Alsatians, or Basque. An individual's feelings about their cultural identity vary. Some have a strong affiliation for the state in which they live, while for others it is to the district, or pays, of their home. But mostly there is a strong affiliation with the ethnic or culture group to which they belong and identify.

Over the past 50 years, an increasing number of people have come to identify with their ethnic background, which as a consequence has led to demands for certain innate rights. These demands, when expressed vigorously, eventually lead to some form of conflict that can be intense. The conflict may be contained domestically, i.e., with the boundaries of a specific country, or become international, spilling over into neighboring states. In the most intense cases, the demands of the people are usually for some sort of political recognition. They want to free themselves from the dominant society in which they live to establish an autonomous region or an independent state. There are many such examples throughout the world; the Uighurs in western China, the Basques in Spain, or Kurds in Turkey, Iran and Iraq. Others want to change the way they are governed to be free from tyrannical leaders as in the Arab uprising of 2011. To truly understand the needs of these culture groups we must try to put ourselves within the culture and experience life as these people undergo daily trials and tribulations. Only through an awareness of the differences and diversities of cultures can we truly understand the reasoning behind some of these conflicts.

An assessment of the conflicts occurring throughout the world will show that there is a more profound historical background than our current perceptions allow. It will be clear that an appreciation of the historical background is just as significant as a grasp of the present-day geography. Thus, in the following pages we will examine the background, development and diffusion of many of the themes. First we ask the question: Why is cultural understanding important?

THE IMPORTANCE OF CULTURAL UNDERSTANDING IN TODAY'S "SHRINKING WORLD"

*So often today, in a world grown smaller, we are forced to confront new things, new people, new cultures new ways of behavior that are different from what we are used to. Change can be scary. People often seek refuge by clinging tightly to what they know and trying to wall out what they do not. Yet when we do that we diminish ourselves. We deprive ourselves of life's richness and at worst we perpetrate ignorance that breeds prejudice and fear.—***Queen Rania of Jordan, 2007**

In the modern world we can email, facebook, or tweet with "friends," people whom we have never physically met, on the other side of the world. Even the cost of making a phone call and speaking with another person halfway around the world has dramatically declined to such an extent that phoning or texting has become commonplace; whereas twenty years ago the cost was so prohibitive that we would think twice before making the call. These examples show that the world in one sense is shrinking and cultures are becoming intertwined.

The process of the "shrinking world" closes the gaps between cultures and their boundaries. Knowledge and understanding of different cultures is now more important than ever. The closer we become and the easier it is to communicate, the more opportunities we have to appreciate one another's culture. On the other hand, with the compression comes the possibility of overlapping, destruction and loss of our diverse cultures.

On a basic level cultural understanding is the ability of an individual to accept, respect and tolerate another culture without causing any turmoil or change within that culture. As the world shrinks, we are now ever closer and more vulnerable to having culturally diverse experiences. Some people find it intriguing and exciting to learn and explore other cultures, while others do not. For the people that do not find cultural exploration exciting, it does not mean that they do not have the ability to have a cultural experience.

With the ease of communication, travel around the world is also becoming easier, more affordable and more widespread, reinforcing the importance of cultural understanding. When one travels to another country with diverse cultural experiences, one must be able to acknowledge, accept and respect the culture one is visiting. If cultural understanding did not exist, or our understanding did not grow, it would be almost be impossible to safely travel and the chances of having a positive experience with the new culture will be limited. Gregorio Billikopf from the University of California states that "As we interact with others of different cultures, there is no good substitute for receptiveness to interpersonal feedback, good observation skills, effective questions and some horse sense. There is much to be gained by observing how people of the same culture interact with each other" (Billikopf, 2009). He also encourages us to not be afraid to ask questions and to not make generalizations. When asking questions about another culture he states not to rely on just one source, thus avoiding generalizations.

To understand a culture one must realize the complexities within a culture. A culture is built upon education, social standing, religion, personality, beliefs and experiences of that particular group (Billikopf, 2009). Culture entails all learned behavior within a group and to only accept one behavior that a culture exhibits does not open our eyes to the entire culture and definitely does not give you an accurate understanding of the culture at hand. In addition, every culture is very distinct and diverse, but all cultures are unified by the basic elements of education, language, beliefs, etc. We must also realize that all these diverse cultures exist in the same world and must function while being intertwined within each other, sharing and trading ideas, while still maintaining their unique customs. Without a general understanding of cultural diversity the world would be very segregated and somewhat chaotic.

With the globalization of business and personal travel there is an increasing opportunity for cultural confusion. Cultural confusion and misunderstandings are the result of interpreting other culture's actions in relationship to your own personal cultural experiences and standards (Crews, 2000). We need to have an open mind to understand a culture. We cannot relate our culture to another culture. We must accept them as completely different in order to reach an understanding of another culture and preserve the diversity. For example, don't think people in a different culture share your beliefs and then become shocked when you discover they have alternate beliefs. Or, when visiting China do not complain that the food offered is Chinese cuisine. The ability to understand a culture does not mean one must assimilate into the culture and live within the culture. One may argue that cultural understanding may decrease cultural diversity and eliminate the intricate cultures, but cultural understanding does not imply that you must adopt and assimilate within new cultures.

CULTURE—SOME BASICS

To writers in newspapers and the popular press, "culture" means the arts (literature, painting, music and the like). To a social scientist, **culture** is the specialized behavioral patterns, understandings, adaptations and social systems that summarize a group of people's learned way of life. Or put another way, **learned social behavior.** In this broader sense, culture is an ever-present part of the regional differences that are the essence of cultural geography. The visible and invisible evidences of culture: buildings and farming patterns, language, political organization and ways of earning a living, are all parts of the spatial diversity that cultural geographers study. Cultural differences over time may present contrasts as great as those between the Stone Age ivory hunters and modern urban Americans. Cultural differences in areas result in human landscapes with variations as subtle as the differing "feel" of urban Paris, Moscow, or New York or as obvious as the sharp contrasts of rural Sri Lanka and the Prairie Provinces of Canada.

Since such tangible and intangible cultural differences exist and have existed in various forms for thousands of years, we need to address the Why, What, Where and How questions. Why, since humankind constitutes a single species, are cultures so varied? What motivated the development of culture attribute? Where were the origins of the different culture regions we now observe? How, from whatever limited areas individual culture traits developed, were they diffused over a wider portion of the globe? How did people who had roughly similar origins come to display significant areal differences in technology, social structure, ideology and the other myriad details and amalgams of geographic diversity? In what ways and why, are there distinctive cultural variations even in "melting pot" societies such as the United States and Canada or in the outwardly homogeneous long-established countries of Europe? Part of the answer to these questions lies in the way separate human groups developed techniques to solve regionally varied problems of securing food, clothing and shelter and, in the process, created regionally distinctive customs and ways of life.

Components of Culture

Culture is transmitted within a society to succeeding generations by imitation, instruction and example. In short, it is learned, not biological; it has nothing to do with instinct or with genes. As members of a social group, individuals acquire integrated and cohesive sets of behavioral patterns, environmental and social perceptions and knowledge of existing technologies. Of necessity, each of us learns the culture in which we are born and reared; but we need not—indeed, cannot—learn its totality. Age, sex, status, or occupation may dictate the aspects of the cultural whole in which an individual becomes fully indoctrinated. A culture, despite overall generalized and identifying characteristics and even an outward appearance of uniformity and conformity, displays a social structure. That is a framework of roles, plus the interrelationships of individuals and established groups. Each individual learns and adheres to the rules and conventions not only of the culture as a whole but also of those specific to the subgroup to which he or she belongs, a subgroup that may have its own recognized social structure.

Culture is an interlocking web of such complexity and pervasiveness that it cannot be grasped and, in fact, may be misunderstood by being glimpsed or generalized from limited, obvious traits. Distinctive eating utensils, the use of gestures, or the ritual of religious ceremony may summarize and characterize a culture for the casual observer. These are, however, only individually insignificant parts of a totality that one can appreciate only when one experiences the whole.

The richness and intricacy of human life compels us to seek those more fundamental cultural variables that give order to societies and to summarize them in meaningful conceptual and spatial terms. We begin with **culture traits,** the smallest distinctive item of culture. Culture traits are units of learned behavior ranging from the language spoken to the tools used or to the games played. A trait may be an object (a fishhook, for example), a technique (weaving and knotting of a fishnet), a belief (in the spirits resident in water bodies), or an attitude (a conviction that fish is superior to other animal protein). Such traits are the most elementary expressions of culture, the building blocks of the complex behavioral patterns of distinctive groups of peoples.

Individual cultural traits that are functionally interrelated comprise a **culture complex.** The existence of such complexes is universal. Keeping cattle was a culture trait of the Masai of Kenya and Tanzania. Related traits included the measurement of personal wealth by the number of cattle owned, a diet containing milk and the blood of cattle and disdain for labor unrelated to herding. The assemblage of these and other related traits yielded a culture complex descriptive of one aspect of Masai society. In exactly analogous ways, religious complexes, business behavior complexes, sports complexes and others can easily be recognized in America or any other society.

Culture traits and complexes have a real extent. When they are plotted on maps, the regional character of the components of culture is revealed. Although human geographers are interested in the spatial distribution of these individual elements of culture, their usual concern is with the **culture region,** a portion of the earth's surface occupied by populations sharing recognizable and distinctive cultural characteristics that summarize their collective attributes or activities. Examples include the political organizations societies devise, the religions they espouse, the form of economy they pursue and even the type of clothing they wear, eating utensils they use, or kind of housing they occupy. There are as many culture regions as there are culture traits and complexes identified for population groups.

INTERACTION OF PEOPLE AND ENVIRONMENT

Culture develops in a physical environment that, in its way, contributes to differences among people. In primitive societies, the acquisition of food, shelter and clothing—all parts of culture—depends on the utilization of the natural resources at hand. The interrelations of people with the environment of a given area, their perceptions and utilization of it and their impact on it are inter-

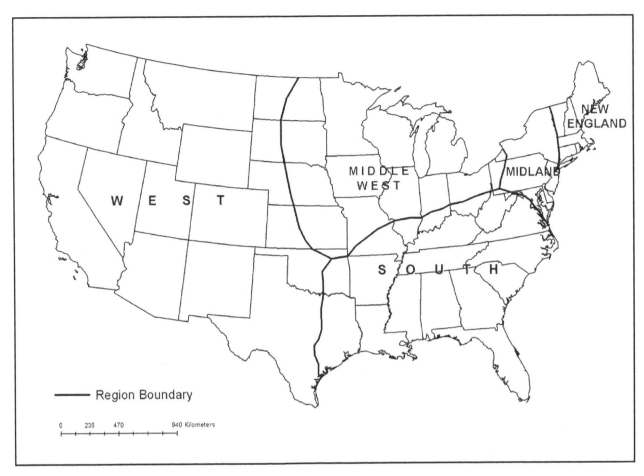

Figure 1.1   Cultural Regions. (University of Akron Produced)

woven themes of **cultural ecology**—the study of the relationship between a culture group and the natural environment it occupies.

Environments as Controls

Geographers have long dismissed as intellectually limiting and demonstrably invalid the ideas of **environmental determinism,** the belief that the physical environment exclusively shapes humans, their actions and their thoughts. Environmental factors alone cannot account for the cultural variations that occur around the world. For instance, levels of technology, systems of organization and ideas about what is true and right have no obvious relationship to environmental circumstances.

The environment does place certain limitations on the human use of territory, although such limitations must be seen not as absolute, enduring restrictions but as relative to technologies, cost considerations, national aspirations and linkages with the larger world. Human choices in the use of landscapes are affected by group perception of the feasibility and desirability of their occupancy and exploitation. These are not circumstances inherent in the land. Mines, factories and cities were (and are being) created in the formerly nearly unpopulated tundra and forests of Siberia as a reflection of Russian developmental programs, not in response to recent environmental improvement. **Possibilism** is the viewpoint that people, not environments, are the dynamic forces of cultural development. The needs, traditions and level of technology of a culture affect how that culture assesses the possibilities of an area and shapes what choices the culture makes regarding them. Each society uses natural resources in accordance with its culture. Changes in a group's technical abilities or objectives bring about changes in its perceptions of the usefulness of the land.

Map evidence suggests the nature of some environmental limitations on use of area. The vast majority of the world's population is differentially concentrated on less than one-half of the earth's land surface. Areas with relatively mild climates that offer a supply of fresh water, fertile soil and abundant mineral resources are densely settled, reflecting in part the different potentials of the land under earlier technologies to support population. Even today, the Polar Regions, high and rugged mountains, deserts and some hot and humid areas contain very few people. If resources for feeding, clothing, or housing ourselves within an area are lacking, or if we do not recognize them there, there is no inducement for people to occupy the territory.

Environments that do contain such resources provide the framework within which a culture operates. Coal, oil and natural gas have been in their present locations throughout human history, but they were of no use to pre-industrial cultures and did not impart any recognized advantage to their sites of occurrence. Not until the Industrial Revolution did these deposits gain importance and come to influence the location of such great industrial complexes as the Midlands in England, the Ruhr in Germany and the steelmaking districts formerly so important in parts of the northeastern United States. American Indians made one use of the environment around Pittsburgh, whereas 19th-century industrialists made quite another.

Human Impacts

Any item constructed by a group for its shelter, comfort, or use can be considered part of the **material culture.** Furthermore, buildings, furniture and tools of a particular group are historical elements that can determine how people lived in the past. They are a valuable source of information about the development of a culture group, their activities and mode of life. Evidence of what the group made, how the artifacts were constructed and to what use they were put are a reflection of man's relationship with the physical environment. People are also able to modify their environment. We modify our environment through the material objects we place on the landscape: cities, farms, roads and so on. The form these take is the product of the kind of culture group in which we live. The **cultural landscape,** is the tangible physical record of a given culture, or as Marsh describes it *the earth's surface as modified by human action* (Marsh, 1864). House types, transportation networks, agricultural activity and the size and distribution of settlements are among the indicators of the use that humans have made of the land.

Human actions, deliberate and inadvertent, modifying and sometimes destroying the environment are perhaps as old as humankind itself. People have used, altered and replaced the vegetation in vast areas of the tropics and midlatitudes. They have hunted to extinction vast herds and whole species of animals and birds, i.e., the Passenger Pigeon of North America. Through overuse and abuse of the earth and its resources, they have rendered sterile formerly productive and attractive regions such as the Northern Sahara.

Fire has been called the first great tool of humans and the impact of its early and continuing use is found on nearly every continent. Polewards of the rain forests of South America, Africa and South Asia lay the tropical savanna—extensive grasslands with scattered trees and groves. Research has concluded that the trees are the remnants of naturally occurring tropical dry forests, thorn forests and scrub that have, over many millennia, been obliterated by the use of fire to remove unproductive forest and to clear off old grasses for more nutritious new growth. These grasses supported vast herds of grazing animals that were the mainstay of the hunter-gatherer societies that inhabited these areas. A modern example of the qualities of the grasslands can be found in Kenya where the government sought to protect its national game preserves by prohibiting the periodic use of fire. However, this turned out to have an undesired effect. The vast grazing herds of gazelles, zebras, antelope, together with the lions and other predators that fed upon them that attracted scores of tourists—a major factor in the country's economy—were being replaced by less-appealing browsing species such as rhinos, hippos and elephants, because with fire prohibited, the forests began to reclaim their natural environment.

The same form of vegetation replacement occurred in midlatitudes. The grasslands of North America were greatly extended by Native Americans, who burned the forest margin to extend grazing areas and to drive animals in the hunt. The control of fire in modern times has resulted in the advance of the forest once again in formerly grassy areas ("parks") of Colorado, northern Arizona and other parts of the western United States.

There is a current romantic notion that primitive cultures "lived in harmony" with their environment. Throughout the spread of human existence on the earth, examples abound of cultural action that has changed nature. During the Stone Age, humans hunted to extinction whole species of large animals on all inhabited continents. This has come to be known as the Pleistocene overkill. Some researchers estimate about 40 percent of Africa's large-animal species passed to extinction, while in North America, most of the original large animal species had disappeared by 10,000 years BP (Before Present) under pressure from the migrating hunters spreading across the continent. It may be that climatic changes were partially responsible, but human action seems the more likely explanation for the abrupt changes in the flora and fauna. It is estimated that the Maoris of New Zealand and the peoples of Micronesia and Polynesia destroyed some 80 to 90 percent of South Pacific bird species by the time Captain Cook arrived in the 18th century.

Not only destruction of animals but of the life-supporting environment itself has been a frequent consequence of human misuse of area. North Africa, the "granary of Rome" during the empire, became wasted and sterile in part because of mismanagement. Roman roads standing high above the surrounding barren wastes, villas with artifacts such as vessels for wine and grain submerged by the engulfing sands of the desert give testimony to the erosive power of wind and water when natural vegetation is unwisely removed and farming techniques are inappropriate. Other examples can be cited: Easter Island in the South Pacific was covered lushly with palms and other trees when Polynesians settled there about A.D. 400. By the beginning of the 18th century, Easter Island had become the barren wasteland it remains today. Deforestation increased soil erosion, removed the supply of timbers needed for the vital dugout fishing canoes and made it impossible to move the massive stone statues that were significant in the islanders' religion. With the loss of livelihood, resources and the collapse of religion, warfare broke out and the population was decimated. A similar tragic sequence is occurring today on Madagascar in the Indian Ocean.

The more technologically advanced and complex the culture, the more apparent is its impact on the environment. In sprawling urban-industrial societies, the cultural landscape has come to outweigh the natural physical environment in its impact on people's daily lives. It interposes itself between "nature" and humans. For example, the residents of the cities of such societies—living and working in climate-controlled buildings, driving to enclosed shopping malls—can go through life with very little contact with or concern about the physical environment.

KEY TERMS

Culture
Culture Traits
Culture Complex
Culture Region
Culture Realm
Cultural Ecology

Environmental Determinism
Possibilism
Cultural Landscape
Material Culture
Relationship

DISCUSSION TOPICS

1. What do you understand Queen Rania's remarks to mean in terms of culture?
2. In an ever changing world, culture diversities become even more important.
3. As the world grows smaller, understanding diverse cultures become more important.

CHAPTER 2

The Development of Cultures

CULTURAL CHANGE

How do cultures begin, expand, grow and alter over time? In this section we will look at the beginnings of a culture, how it diffuses, changes and assimilates other cultures.

Neolithic man is reputed to be the earliest pastoralist, but this does not mean that culture began then. Palaeolithic cultures probably had a social order even if this was limited to the men hunting and the woman looking after the homestead. Every human population has needed to evaluate the economic potential of its inhabited area and has organized its life about its natural environment in terms of the skills available to it. Wherever men live, they have operated to alter the aspect of the Earth for their own well-being (Sauer, 1971).

The recurring theme in the study of cultures is change. It is argued that no culture is or has been characterized by a permanently fixed set of material objects, systems of organization, or ideologies. While culture's are essentially conservative, they're always in a state of flux. Some changes are major and pervasive. The transition from hunter-gatherer to sedentary farmer affected every facet of the cultures experience. A profound impact has been the Industrial Revolution, which with its associated rise in urbanization, has impacted all societies it has touched.

Change may be so slight as to go almost unnoticed, or it may be extensive and impact large societies such as the agricultural revolution or the *Industrial Revolution* of the 17th, 18th and early 19th centuries. The spread of culture is most easily investigated via particular cultural traits, though difficulties do arise in that an entire band of cultural relations may be welded together by a particular culture group and spread over a wide area from the central region. According to Vidal de la Blanche a "Genre de Vie" or "way of life" may be traced from its inception to its expansion into a greater territorial base and finally to the point where it is absorbed or replaced by another expanding culture.

But where does this all begin? Geographers recognize a specific location for the genesis of a trait, an idea, or an innovation from which it spreads across continents and eventually the world.

Cultural Hearths

The term **cultural hearth** is used to describe these centers of innovation and invention from which key culture traits and elements moved to exert an influence on surrounding regions.

Why hearth?—A logical question students pose when first introduced to the term. Think then of the earliest Neolithic cultures coming in from the hunt and gathering around a fire, a place for warmth and cooking. They might discuss the day and may share experiences. Some may have found new ways of trapping or met a group from another tribe who had different weapons. These ideas and innovations spread from the fireside, also called a **hearth**—the place where the fire is built—to others in the camp or tribe and from there outward. Cultural Geographers have taken this term and related it to the beginning of any cultural trait, such as agriculture, writing, language and beliefs. It may also be viewed as the cradle of a cultural lifestyle.

Research shows that many hearths have existed, but only a few have evolved across the world with the social and technical development that can be called **civilization.** Civilization is an imprecise term that generally reflects the achievements of a group of people, commonly assumed to include writing, astronomy and mathematics, working with metals, trade, a formalized governmental system and an urban culture.

Several major cultural hearths emerged in the Neolithic period. Prominent centers of early creativity were found in Egypt, Crete, Mesopotamia, the Indus Valley of the Indian subcontinent, northern China, southeastern Asia, several locations in sub-Saharan Africa, in the Americas and elsewhere. They arose in widely separated areas of the world at different times and under differing ecological circumstances. Each displayed its own unique mix of culture traits.

Each cultural hearth showed a rigorous organization of agriculture resulting in local productivity sufficient to enable a significant number of people to engage in nonfarm activities. Therefore, each hearth region saw the creation of a stratified society that included artisans, warriors, merchants, scholars, priests and administrators. Each also developed or adopted astronomy, mathematics and the all-essential calendar. Each, while advancing in cultural diversity and complexity, exported technologies, skills and learned behaviors far beyond its own boundaries.

As peoples settled down, they developed technologies that made the new sedentary life more comfortable. They developed the art of spinning and weaving plant and animal fibers for clothing and household uses. They learned to make pots and to fire clay and make utensils. Later, they developed brick making and construction skills. Instead of hunters or pastoralists, there developed a whole new structure within the culture—specialist occupations such as metalworkers, potters, merchants and scribes, who tended to the needs of the farmers and hunters.

A prime example of civilization and one that has been recognized as an indicator of a culture's advancement is writing. **Writing** as a means of keeping tally on one's property (how many goats, camels, etc. were owned) and keeping tally of trading activities (what was sold and bought and for how much), appeared first in Mesopotamia as cuneiform and in Egypt as hieroglyphics at least 5,000 years ago. The separate forms of writing have suggested to some that they arose independently in separate hearths. However, a second school of thought suggests that writing originated in Mesopotamia, then spread outward to Egypt, to the Indus Valley, to Crete and perhaps even to China, though it is usually assumed that Chinese ideographic writing developed independently. Systems of record keeping developed independently in the New World. Once created, these writings spread widely in areas under the influence of Andean and Mesoamerican hearths between 2,600 and 2,300 years ago.

With new technologies, an increasingly explorative culture appears alongside a more formal economy. People start to move across the earth freely trading their surplus produce and artifacts, exchanging ideas and innovations; all the while they are assimilating new ideas and innovations; changing and developing their cultural perspective.

THE PROCESS OF CULTURAL CHANGE

In the previous section we questioned how cultures begin, expand, grow and alter over time. In that section, we examined the development of the cultural hearth and various forms of cultural society. Now, let's consider how cultures change, develop and grow. Think of how the culture of the United States, as it is today, has changed over the last century. Think of all the electrical, electronic and transportation devices introduced in that time period. Consider, also, the social behavior and recreational changes that have occurred. Included in these changes will be our greater reliance on electronic gadgetry, changes in employment patterns and attitudes toward the role of women in society. Such changes occur because the cultural traits of any group are not independent. They are clustered in a coherent and integrated pattern. Small-scale changes may have wide repercussions if they are accommodated and adopted by an associated society. Changes within and between cultures are induced by **innovation, diffusion and acculturation.**

Innovation

Innovation implies changes to a culture that result from ideas created within the social group itself and adopted by the culture.

Many innovations are of little consequence on their own, but the widespread adoption of something seemingly inconsequential may bring about large changes when viewed over time. For example, a song or piece of music may lead many individuals to favor that tune, which could bring about new dance routines, which could in turn lead to changes in fashion, which would impact retailers, advertisers and consumption patterns. Eventually, a new cultural form will be identified. That may have an important impact on the thinking processes of the adopters and those who come into contact with them. An historic example of all the aforementioned cultural impacts and innovation occurred in the 1960s after The Beatles released their "Sgt. Pepper's Lonely Hearts Club Band" album. This impacted many facets of culture; dress, marketing, social attitudes, etc.

Society has an innate resistance to change since innovation inevitably creates tensions between the new reality and the older established socio-economic conditions. Tensions are usually only solved by adaptive changes in the society at large. The gap that may develop between the newly adopted technology and a slower paced, older social trait is called a **cultural lag.** In modern society, innovative change has become common, expected and inevitable. The rate of invention has increased and the period between an idea's conception and product availability has become smaller. The spatial implication is that large urban centers tend to be the focal point of innovation. Ideas not only stimulate new thoughts, but also create circumstances in which society must develop new solutions to maintain its forward momentum.

Culture Shock

In October 1957, America experienced a major cultural shift; a shock that resulted in a distinct cultural change and development. This shock was caused by a Russian satellite that was orbiting the earth. Sputnik, the first manmade object to circle the planet, caused a major change in the American culture. It was the process that developed the high-tech culture that Americans enjoy today.

A brief examination of these sudden changes, their cause and long-term effects will show that a group can undergo trauma, survive and develop into a new form. Is cultural development a gradual, ongoing process or do humans need a shock to awaken the cultural process? The impact of the shock may initially be slight, or as with the Acadians (see later section), it can be dramatic. This shock can occur in many forms and in different spheres of the societal body. It may be an endogenous or an exogenous occurrence.

Had Sputnik not given the American psyche a jolt, it is likely that American technology would have developed its own satellites, but then the shock would have been endogenous and less severe because it was generated by the culture.

CULTURAL DIFFUSION

How do ideas spread within the culture group and more importantly to those outside?

The anthropologist Julian Steward (1902–1972) proposed the concept of multilinear evolution to explain how widely separated cultures developed under similar ecological circumstances. He suggested that each major environmental zone—arid, high altitude, midlatitude steppe, tropical forest—tends to induce common adaptive traits in the cultures of those who live there. Those traits were founded on the development of agriculture and the emergence of similar cultural and administrative structures in the several cultural hearths as described in the previous section. But, *similar* does not imply *identical*. Steward simply suggested that since comparable sequences of developmental events cannot always or even often be explained on the basis of borrowing or exporting ideas and techniques (because of time and space differences in cultures sharing them), they must be regarded as evidence of parallel creations based on similar ecologies. From similar origins, but through separate adaptations and innovations, distinctive cultures emerged.

Contrary to Steward's belief is the idea that cultural similarities occur primarily by spatial spread—diffusion from one or more common origin sites. Cultural advancement and civilizations are passed on along trade routes and through group contact rather than being the result of separate and independent creation. Although long out of favor, diffusionism has recently received renewed support from archaeological discoveries apparently documenting very long-distance transfer of ideas, technologies and language by migrating peoples.

The common characteristics, derived from multilinear evolution and the spread of specific culture traits and complexes, contained the roots of **cultural convergence.** This term describes the sharing of technologies, organizational structures, traits, ideas and artifacts that are evident among widely separated societies, which in a modern world are united by instantaneous communication and efficient transportation. Convergence in these worldwide terms is, for many observers, proof of the pervasive globalization of culture.

Spatial Diffusion

The process by which an idea or innovation is transmitted from one individual or group to another across space is called **spatial diffusion.** This diffusion process may assume a variety of forms, each different in its impact on social groups. People move for any of a number of reasons to a new area and take their culture with them. For example, immigrants to the American colonies brought along crops and farming techniques, building styles, or concepts of government alien to their new home. In other instances, information about an innovation (e.g., hybrid corn or compact discs) may spread throughout a society, perhaps aided by local or mass media advertising or new adopters of an ideology or way of life.

The terms we use for these forms of diffusion are *expansion diffusion* and *relocation diffusion*.

Expansion Diffusion

We will begin the discussion with **expansion diffusion.** This involves the spread of an item or idea from one place to others. In the process, the thing diffused also remains—and is frequently intensified—in the origin area. Islam, for example, expanded from its Arabian Peninsula origin locale across much of Asia and North Africa. At the same time, it strengthened its hold over its Near Eastern birthplace by displacing pagan, Christian and Jewish populations. We can sub-divide expansion diffusion into further categories:

1. Contagious Diffusion

When an innovation, idea or other cultural trait spreads outward from its inception area in a fairly uniform manner we call this **contagious diffusion.** The term implies direct contact between those who developed the idea—**the innovators**—or have adopted the innovation—**the adopters**—and those who newly encounter it—**the receptors.** In the field of medicine and medical geography, this form of diffusion relates to infectious diseases. If an idea has merit in the eyes of potential adopters and they themselves become adopters, the number of contacts of adopters with potential adopters will compound. Consequently, the innovation will spread slowly at first and then more and more rapidly until saturation occurs or a barrier is reached.

The rate of diffusion of a trait or innovation may be influenced by **time-distance decay,** which simply tells us that the spread or acceptance of an innovation is usually delayed as distance from the source of the innovation increases. In some instances, however, geographic distance is less important in the transfer of ideas than is communication between major centers or important people. News of new clothing styles, for example, quickly spreads internationally between major cities and only later filters down irregularly to smaller towns and rural areas.

2. Hierarchical Diffusion

The process of transferring innovations, first between larger places or prominent people and only later to smaller or less important points or people, is known as **hierarchical diffusion.** The Christian faith in Europe, for example, spread from Rome as the principal center to provincial capitals and thence to smaller Roman settlements in largely pagan occupied territories. Today, new discoveries are shared among scientists at leading universities before they appear in textbooks or become general knowledge through the public press. The process works because, for many things, distance is relative to the communication network involved. Big cities or leading scientists, connected by strong information flows, are "closer" than their simple distance separation suggests.

While the diffusion of innovations may be slowed by time-distance decay, their speed of spread may be increased to the point of becoming instantaneous through the space-time compression made possible by modern communication. Given access to radios; telephones; worldwide transmission of television news, sports and entertainment programs; and—perhaps most importantly—to computers and the internet, people and areas distantly separated can immediately share in a common fund of thought and innovation. Modern communication technology has encouraged and facilitated the globalization of culture.

3. Stimulus Diffusion

The third form of expansion diffusion summarizes situations in which a fundamental idea, though not the specific trait itself, stimulates imitative behavior within a receptive population. A documented case in point involves the spread of the concept, but not of a specific system of writing from

European American settlers to at least one Native American culture group. Around 1820, Sequoyah, a Cherokee who could neither read nor write any language, by observing that white people could make marks on pieces of paper to record agreements and repeat lengthy speeches, devised a system for writing the Cherokee language. He eventually refined his initial complex pictorial system to a set of 86 syllabic signs. With time, literacy in the new system spread to others and *The Cherokee Phoenix*, a Cherokee language newspaper, was established in 1828. There was no transfer between cultures or groups of a specific technique of writing, but there was a clear-cut case of the idea of writing diffusing by stimulating imitative behavior.

Relocation Diffusion

In relocation diffusion, migrating individuals or populations that possess an innovation or idea carry it to a new area. **Mentifacts** (the central elements of a culture—beliefs, language, etc.) and/or **artifacts** (the physical materials of a culture—tools, buildings, clothing and the like) are therefore introduced and assimilated into new locales by new settlers who become part of the new population. The spread of religions by settlers or conquerors is a clear example of relocation diffusion, as was the diffusion of agriculture to Europe from the Middle East. Europeans brought their faiths to areas of colonization or economic penetration throughout the world. On the world scale, massive relocation diffusion resulted from the European colonization and economic penetration that began with the 16th-century 'voyages of discovery.' More localized relocation diffusion continues today as Asian refugees or foreign "guest workers" bring their cultural traits to their new areas of settlement in Europe or North America.

Innovations in technology and ideas may be relatively readily diffused to and accepted by, cultures that have basic similarities and compatibilities. For example, the innovations of the Industrial Revolution were easily and quickly adopted by Continental Europe and North America as they diffused from England with whom they shared a common economic and technological background. Industrialization was not quickly accepted in Asia and Africa where societies had totally different cultural backgrounds. On the ideological level, successful diffusion depends on acceptability of the innovations. The Shah of Iran's attempt at rapid westernization of traditional Iranian, Islamic culture after World War II eventually provoked a nationalist backlash that led to a revolution that deposed the Shah and established clerical control of the state.

Diffusion is not solely the outcome of knowledge dispersal, but must include the acceptance of new traits, articles, ways of doing and/or thinking by the potential receiving population. This depends not just on information flow to that population, but also upon its entire cultural and economic structure. Innovations may be rejected not because of lack of knowledge, but because the new trait violates the established cultural norms of the culture to which it is introduced.

For example, a well-intentioned development organization may suggest to a peasant agricultural society that they develop a cash crop. This idea may be rejected not because it is not understood, but because it unacceptably disrupts the knowledge base and culture complex devoted to an assured food supply that the subsistence farming economy provides. Similarly, less disruptive new production ideas introduced into agricultural societies, such as chemical fertilizers, deep-well irrigation and hybrid seeds, may be rejected simply because, though understood, they are not affordable. A culture is a complex organized system. Cultural change involves alteration of the system's established structure in ways that may be rejected even after knowledge of an innovation is received and understood.

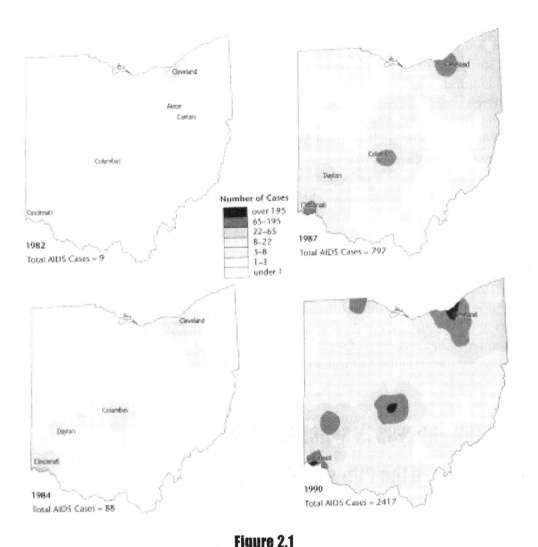

Figure 2.1

Gould, Peter 1993 *The Slow Plague: A Geography of the Aids Pandemic*, Cambridge, Blackwell, pgs. 68–69.
ISBN 978-1-55786-418-7 Reprinted by permission.

Study these maps of Ohio. They show the diffusion of AIDS throughout the state during the 1980s.

What types of diffusion is taking place?

Can you associate any significance to the dispersion?

Acculturation

A culture group may undergo major modifications in its own identifying traits by adopting some or all of the characteristics of another, more dominant, culture group. Immigrant populations take on the values, attitudes, customs and speech of the receiving society, which itself may undergo certain changes resulting from absorbing the migrant group.

A different form of contact and subsequent cultural alteration may occur in a conquered or colonized region where the subordinate or subject population is either forced to adopt the culture of the new ruling group, introduced through relocation diffusion, or does so voluntarily; overwhelmed by the superiority in numbers or the technical level of the conqueror. Tribal Europeans in areas of Roman conquest, native populations in the wake of Slavic occupation of Siberia and Native Americans stripped of their lands following European settlement of North America experienced this kind of cultural modification or adoption.

In extreme cases, of course, small and particularly primitive indigenous groups brought into contact with conquering or absorbing societies may simply cease to exist as separate cultural entities. Although presumably such cultural loss has been part of all of human history, its occurrence has been noted and its pace quickened over the past 500 years. By one informed estimate, at least one-third of the world's inventory of human cultures has totally disappeared since A.D. 1500, along with their languages, traditions, ways of life and, indeed, with their very identity or remembrance.

In many instances, close contact between two different groups may involve adjustments of the original cultural patterns of both rather than the disappearance of either. For example, after World War II, occupying American forces imposed changes in Japanese political organization and philosophy. In addition, the Japanese voluntarily adopted some more frivolous aspects of American life. In turn, American society was enriched by the selective importation of Japanese cuisine, architecture and philosophy, demonstrating the two-way nature of cultural diffusion. Where that two-way flow reflects a more equal exchange of cultural outlooks and ways of life, a process of **transculturation** has occurred. That process is observable within the United States as massive South and Central American immigration begins to intertwine formerly contrasting cultures and brings about alterations to them both.

THE DIFFUSION PROCESS IN SPACE AND TIME

Cultural Shock and the Diffusion Process

If we accept that diffusion is the way cultures develop and progress, we can use the analogy of the ripples on a pond caused by a stone that has been thrown into the center of the pond, breaking the surface of the water. Supposing these ripples are radiating from the center of the pond and all of a sudden another stone is dropped in, halfway between the center and the shore, the original ripples will be affected in proportion to the size of the second stone. If the second stone has a greater mass than the first, the affect on the pond is dramatic. Putting this analogy into cultural terms, we use the evidence of the impact of European cultures on the aboriginal Americans which was profound and dramatic.

If the second stone is much smaller than the first, there is very little change in the pattern of events; the original culture undergoes slight change. For example, the impact of Europeans in China was fairly nondescript.

If the two pebbles are of similar size, then there is an initial cultural shock, but over time there is a sharing of ideas and each culture will continue having absorbed a degree of the other. The closest example here is the Indian British cultural exchange of the 19th century. Although it will be argued that the British impact on India was greater, there was still a contra flow from the Indian culture to the British. This is evidenced in the number of words of Indian origin that have assimilated into the language and, to a lesser degree, the artifacts that have also been absorbed by both sides.

Diffusion Waves

Much geographic interest in diffusion studies stems from the work of the Swedish geographer Torsten Hägerstrand and his colleagues at the University of Lund. Hägerstrand's *Innovation Diffusion as a Spatial Process* was originally published in Sweden in 1953. It was concerned with the spread of several agricultural innovations, such as bovine tuberculosis controls and subsidies for the improvement of grazing in an area of central Sweden (Hägerstrand, 1953). This book was the precursor of various practical studies, particularly in the United States.

In one of his early studies of the diffusion process, Hägerstrand suggested a four-stage model for the passage of what he terms *innovation waves*, but which are more generally called diffusion waves. Hägerstrand created maps of the diffusion of various innovations in Sweden, ranging from bus routes to agricultural methods.

More advanced work on the shape of diffusions in space and time has confirmed that diffusion occurs in essentially three-dimensional wavelike forms. By fitting generalized contour maps (called *trend* surface maps) to the original Swedish data, American geographer Richard Morrill (1970) showed that a diffusion wave has height, reflecting the rate of acceptance. At first it has limited height signifying few acceptors, but over time it increases in both height and extent and then decreases in height but increases in total area covered. The gradual weakening of the wave over time and space is both time-dependent and space-dependent. One feature that comes to light is that a wave travelling from one center of innovation will lose height and strength when it meets a wave coming from another direction.

Distance Decay and Time Delay

Another way to describe the diffusion process is to examine the effect that distance from the originating source has on the acceptance of an innovation. A person at some distance from an innovator is less likely to accept the innovation than someone close by. Consequently, the impact of the innovation is less or decays with distance. This is clearly shown in the Hägerstrand Model outlined in the next section.

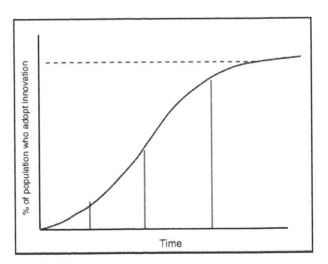

Figure 2.2   Time Distance Decay.

Another aspect related to the diffusion of innovations is the effect of time. It should be realized that an adopter at some distance from the source of the innovation will also be affected by time. However, there is a second aspect of time to consider; this is related to the speed of adoption by a group which can be called the impact of time delay, which does not factor in distance. People close by the innovation may delay in their adoption for reasons other than distance. An example of the impact of time delay in the diffusion process is explained by Grigg in relation to agricultural activity. He suggests that an adoption curve over time shows a slow rate of adoption at first, then accelerates as more people are adopters and slows when only the laggards are left. Shown in a characteristic S curve, this is a function of the rate of the flow of information. Innovators and early adopters hear of the innovation on the radio, from agricultural extension workers, or agents selling new seed or fertilizers; late adopters hear of the innovation only from their neighbors' and adopt only after watching their neighbors' experience (Grigg, 1974).

Using the above examples we can establish four stages in the diffusion process, each describing the passage of an *adoption curve* or *wave through* an area.

1. The *primary stage* is the stage in which centers of adoption are established. These we can call the hearths of the innovation.
2. The *diffusion stage* is the start of the actual diffusion process; there is movement away from the hearth area in many forms of the spatial diffusion process. As the stage continues new centers of innovation in distant areas are established.
3. The *condensing stage* involves the relative increase in the acceptance of an innovation—the time delay. This is equal in all locations, regardless of their distance from the hearth.
4. In the final or *saturation stage* there is marked slowing and eventual cessation of the diffusion process. The innovation has been accepted throughout the entire area, with one or two laggards or non receptors. The area returns to a modified homogeneous surface as it was before the process began.

Following Hägerstrand's initial work, other Swedish geographers carried out studies to test the validity of this four-stage process. Gunnar Tornqvist, for instance, traced the spread of televisions in Sweden by observing the growth of TV ownership from an origin in 1956. Using data from 4,000 Swedish post office districts, he demonstrated that television was introduced into Sweden relatively late, yet within nine years about 70 percent of the country's households had bought their first set. The result of the Swedish and other geographers' studies confirm Hägerstrand's original analysis and shows that innovation diffusion follows the four-stage pattern.

The Basic Hägerstrand Model

From his empirical studies Hägerstrand went on to suggest how a general operational model of the process of diffusion could be built. There follows a description of how the simplest of the **Hägerstrand models** was constructed.

Contact Fields

By looking at examples of spatial diffusion we have studied, we see that the probability that an innovation will spread is related to distance. If we measure the distance between two possible innovators we can then relate this variable to the spread of information through a population.

First we make the assumption that the probability of contact between any two people (or groups, or regions) will get weaker the further apart they are. If we call one person the **sender,** we can say that probability of any other person (call them the **receptor**) receiving a message from the sender is inversely proportional to the distance between them. Near the sender, the probability of contact will be strong, but will become progressively weaker as the distance from the sender increases. The exact form of decline with distance is difficult to judge, but the evidence from previous studies indicates that it may be exponential. That is, it may fall off steeply at first but then even out.

This spatial pattern is called a **contact field.** In diffusion models we can retain the exponential contact field but replace geographic distance with economic distance between cities in an urban hierarchy or with social distance in a social-class hierarchy. Distance may not be symmetric in the hierarchic cases; for example, population migration up the hierarchy (from small to large towns) may be easier than migration down the hierarchy. This implies that the socio-economic distance between levels depends on the direction of movement.

The contact fields in epidemics may be very complex. For instance, studies of measles indicate that the probabilities of contact (and thus infection) within a given group such as a family or the students at a school may be random. However, the probability of contact *between* such groups may be exponentially related to distance in the way we have already described. For example, research on southwest England showed that the probability of measles outbreaks in an area immediately adjacent to an area that had already reported cases was about one in eight. With further distance from the infected area the probabilities of infection fell steadily to about one in thirty.

MEAN INFORMATION FIELDS

To translate the general idea of a contact field into an operational model that can be used to predict future patterns of diffusion, Hägerstrand used probabilities of contact to determine a **mean information field (MIF),** that is, an area, in which contacts could occur. Superimposing the field on a square grid of 25 cells enabled him to assign each cell a probability of being contacted. The probability (P) of contact for the central cells is very high, in fact, over 40 percent ($P = 0.4432$). For the corner cells at the greatest distance from the center, the probability of contact is less than 1 percent ($P = 0.0096$).

0.0096	0.0140	0.0168	0.0140	0.0096
0.0140	0.0301	0.0547	0.0301	0.0140
0.0168	0.0547	0.4432	0.0547	0.0168
0.0140	0.0301	0.0547	0.0301	0.0140
0.0096	0.0140	0.0168	0.0140	0.0096

TABLE 2.1   Probability of Hits.

0–95	96–235	236–403	404–543	544–639
640–779	780–1080	1081–1672	1628–1928	1929–2086
2069–2236	2237–2783	2784–7215	7216–7762	7763–7930
7931–8070	8071–8371	8372–8918	8919–9219	9220–9359
9360–9455	9456–9595	9596–9763	9764–9903	9904–9999

TABLE 2.2   Mean Information Field.

To make the grid operational, a **Monte Carlo model** is used. The upper left cell is assigned the first 96 digits within the range 0 to 95; the next cell in the top row has a higher probability of contact ($P = 0.0140$) and is assigned the next 140 digits within the range 96 to 235 and so on. Continuing the process gives the last cell the digits 9903–9999, to make a total of 10,000 for the complete MIF. These numbers are important in "steering" messages through our simple distribution of population.

Rules for the Basic Hägerstrand Model

The rules given here refer only to the simplest version. They can be relaxed to allow modifications and improvements.

1. Assume that the area over which the diffusion takes place consists of a uniform plain divided into a regular set of cells with an even population distribution of one person per cell.
2. Time intervals are discrete units of equal duration (with the origin of the diffusion set at time t_o). Each interval is termed a generation.
3. Cells with a message (termed "sources" or "transmitters") are specified or "seeded" for time *to*. For instance, a single cell may be given the original message. This provides the starting conditions for the diffusion.
4. Source cells transmit information once in each discrete time period.
5. Transmission is by contact between two cells only; no general or mass media diffusion is considered.
6. The probability of other cells receiving the information from a source cell is related to the distance between them.
7. Adoption takes place after a single message has been received. A cell receives a message in time generation t_x from the source cell and, in line with rule 4, transmits the message from time t_{x+1} onward.
8. Messages received by cells that have already adopted the item are considered redundant and have no effect on the situation.
9. Messages received by cells outside the boundaries of the study area are considered lost and have no effect on the situation.
10. In each time interval a mean information field (MIF) is centered over *each* source cell in turn.
11. Location of a cell within the MIF to which a message will be transmitted by the source cell is determined randomly, or by chance.
12. Diffusion can be terminated at any stage. However, once each cell within the boundaries of the study area has received the message, there will be no further change in the situation and the diffusion process will be complete.

The model was originally set up to simulate the spread of information. It could equally well be used to model the spread of infections in a population; in that case the MIF would be a "mean infection field."

Applying the Model

In order to use the model, the basic codes need to be understood. These are set out in "Rules of the Hägerstrand model". The key to putting this model into use is in rules 10 and 11. In each time interval the MIF is placed over *each* source cell so that the center cell of the grid corresponds with the source cell. A *random number* between 0000 and 9,999 is drawn and used to direct the message, following rules 4 to 6. *Random numbers* are sets of numbers drawn purely by "chance" (e.g., by rolling a dice). They can be taken from published tables of random numbers, or generated on a computer, or, for small problems, drawn from a hat. This process is seen in the figures below.

The original adopter is decided upon and the cell in which it resides is the center of the mean information field. This is designated as time 0 (t_0).

In the first generation the number 0624 is drawn from a table of random numbers and a message is passed to a cell that lies to the northeast of the original adopter, located in the source cell (t_1).

In the second iteration, the original adopter (OA) and the first adopter (0624) become transmitters (o). A set of random numbers are generated using each point as the center of the mean information field.

In the next iteration one selection lands in the middle of the MIF (x) and is not counted as a hit. It is not an adopter; therefore, another generation of random numbers takes place using the originating transmitter.

These figures represent the first few stages in the diffusion process. In each generation the MIF is re-centered in turn over each cell that has the message.

Because the Hägerstrand model uses a random mechanism, each experiment or trial produces a slightly different geographic pattern. If we ran thousands of such trials, using a computer,

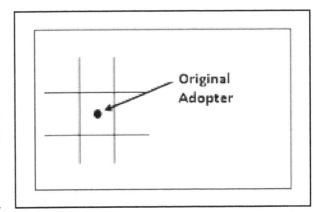

Figure 2.3   Hägerstrand Model Generations.

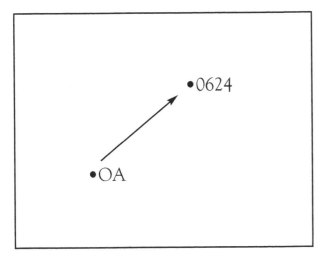

Figure 2.4   First Generation.

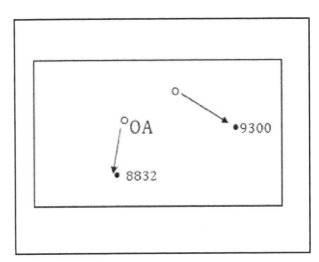

Figure 2.5   Second Generation (t_2).

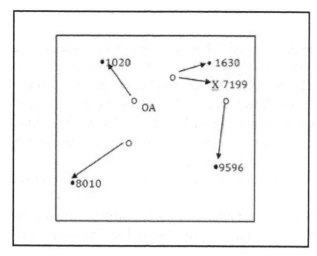

Figure 2.6 Third Generation (t₃).

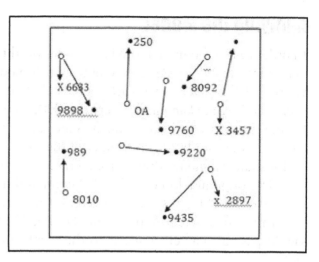

Figure 2.7 Fourth Generation (t₄).

we would find that the sum of all the different results matched the probability distribution in the original MIF; that is, we should arrive back at our starting distribution. It is best if the model is applied to complicated, unpredictable diffusions whose end result is in doubt rather than a simple, predictable diffusion whose end result is known.

KEY TERMS

Cultural Hearth	Contagious Diffusion	Innovation Waves
Cultural Shock	The Innovators	Condensing Stage
Civilization	The Receptors	Saturation Stage
Innovation	Time-Distance Decay	Hägerstrand Models
Diffusion	Hierarchical Diffusion	Sender
Acculturation	Stimulus Diffusion	Receptor
Cultural Lag	Relocation Diffusion	Exponential
Cultural Convergence	Mentifacts	Contact Field
Spatial Diffusion	Artifacts	Mean Information Field
Expansion Diffusion	Transculturation	Monte Carlo Model

DISCUSSION TOPIC

In the modern world, electronic communications has impacted the notion of distance decay!

CHAPTER 3

Cultural Reflections on the Landscapes

"Culture is the agent; the natural area is the medium. The cultural landscape the result."—**Sauer, 1925**

In chapter one we defined the cultural landscape as "the earth's surface as modified by human action," an idea put forward in the 19th century. Later in the 20th century, Carl Sauer (1925) suggested that "A cultural landscape is fashioned from a natural landscape by a culture group." **Cultural landscapes** are complex phenomena. It has been accepted for some time that landscapes reflect human activity and are infused with a variety of cultural values. Jackson tells us that landscape "is *never* simply a natural space, a feature of the natural environment but is the place where we establish our own human organization of space and time" (1984).

Like much of the **natural environment,** cultural landscapes are dynamic. They have evolved over time. As human activity has changed, so too has the landscape. Cultural landscapes reflect political as well as social and cultural paradigms; they are in a continuous process of development. The landscape is a cultural expression that does not happen by chance but is created either informally or intentionally over time; it develops as the culture and natural environment changes. The character of the landscape thus reflects the values of the people who have sculpted it and who continue to live in it. The cultural landscape then is subject to change either by the development of a culture or by a replacement of a culture (Sauer, 1971).

LANDSCAPES AND CONCEPTS

We see landscape through our existing mindsets, influenced in part by what we already know or expect of our concepts of culture. Modern aspects of society, such as television, movies, travel and tourism and the internet build within each individual, a sense of place. Cultural landscapes are influenced in part by what we already know or expect and by the things which interest us: history, food and wine, visual arts and so on. These different mindsets have a profound influence on the way we "experience" places.

Although the culture itself is the shaping force of the landscape, humankind has developed a mindset as to the location of a particular landscape based on preconceived notions of a culture and its associated geography. For example, examine the picture on the next page. There are several cultural aspects that give us a clue to the location.

Jackson (1984) has suggested that the "commonplace aspects of the contemporary landscape, the streets and houses and fields and places of work" can tell us a great deal about a society. These **vernacular landscapes,** or "landscapes of the everyday" are fluid and "identified with local custom, pragmatic adaptation to circumstances and unpredictable mobility."

Figure 3.1   (© 2011, Natali Glado. Used under license from shutterstock.com)

What cultural impacts do you see in this picture?

1 _____

2 _____

3 _____

4 _____

5 _____

Figure 3.2 (© 2011, Lu Wenjuan. Used under license from Shutterstock, Inc.)

Here is a vernacular landscape of streets, houses, high-rise office buildings and even a park. But what makes this landscape so distinctive? Where is it?
Or this one?

Figure 3.3 (© 2011, Leonard Peacefull)

Modes of the Cultural Landscape

The cultural landscape is a record of humankind's occupancy on the natural environment. We can see many modes of this imprint. There are the **modes of production** or land utilization, agricultural activity, mining and forestry. There are the **modes of communication,** roads, railways, canals, ports and airports. There are the **material modes** such as the signage along the highways, the street furniture—lamp posts, seats and resting places—and in certain parts, public conveniences. And finally there are the **built modes:** houses, offices, factories, barns, the enclosures built for protection and walls and fences. The latter are recognized as the built environment or built landscape.

Figure 3.4   Little Langdale. (© 2011, Leonard Peacefull)

This picture illustrates some of the forms of the cultural landscape.
     Using the list below, associate each mode with an item in the picture.

1 Production mode _____

2 Mode of communication _____

3 Material mode _____

4 Built mode _____

THE BUILT ENVIRONMENT

In this section we will first consider that segment of man's physical environment which is purposely shaped by him according to culturally dictated plans (Deetz, 1977). The **built environment** (or landscape) is a term coined by sociologists and favored by planners, engineers and landscape architects (Noble, 2007). It has recently found prominence within the field and can be used synonymously with **material culture,** the folk landscape or settlement landscape.

Nothing reveals more about a culture than the buildings they construct for shelter, economic support, defense and worship. As we reflect on the cultural landscape we must keep in mind that under the influences of a given culture the landscape undergoes development passing through several phases until it reaches the end of its cycle. Then, "with the introduction of a new or alien culture rejuvenation sets in and a new landscape is superimposed on the older one" (Sauer, 1925). In terms of the built environment we often see a juxtaposition of the older culture with the new. This is mostly true in the so-called Old World—Europe, the Middle East and Asia. Take for instance the city of Cairo, Egypt, where the ancient Egyptian cultural landscape at Giza epitomized by the Great Pyramids and the Sphinx, built nearly 5,000 years ago (2560 BCE), overshadows the new high-rise development of modern Cairo.

Another example is in the city of London where modern buildings sit side by side with the remains of the Roman city that are over 2,000 years old.

Although we will use the term built environment to describe all forms of cultural buildings, there are some scholars who prefer to use a variety of terms to describe the less formal architectural styles. In his work on Traditional Buildings, Allen Noble (2007) notes the different terms used: traditional building/architecture, vernacular architecture, or ethnic architecture as opposed to **formal architecture,** which is viewed by professional architects as an art form. He defines traditional building as "using procedures and material objects whose elements are passed from generation to generation, usually orally. . . . This is not to imply that traditional processes do not change over time. They often do." (Noble, 2007).

Figure 3.5   Roman Wall—London. (© 2011, Leonard Peacefull)

Houses are the most significant of cultural indicators, defining areas of cultural influence. The design of cultural dwellings is often strongly influenced by the environment. In traditional cultures or societies, people have to make do with whatever is at hand. The shape and arrangement of buildings is constrained by whatever building materials are available; for example, in parts of northern Europe the local stone is the material of choice but in tropical areas mud and sticks are often the only available material. Also in many cultures around the world, the dwelling building is not necessarily as important as it might be in a western culture. The land on which the dwelling stands is more important because there are the enclosures for the livestock and agricultural produce that create the livelihood of the family. In South Asia, ownership of land is more important than ownership of a house and this can also be reflected in East African cultures where corralling of livestock is of utmost importance as the number of cattle signifies the wealth of their owner.

In geographic studies of the cultural landscape many and varied terms are frequently encountered to describe the type and form of particular cultural buildings. The term **Vernacular Architecture** is taken to mean native or indigenous buildings that are built according to local custom and are not brought in by outsiders, while **Ethnic Architecture** relates buildings introduced by culture groups that have diffused into a region. The new arrivals will probably use different materials, their style of buildings and method of construction will relate to their culture. Ultimately over time, the older vernacular buildings may conform to the new ethnic styles as innovations are taken on board. Thus, the traditional building patterns develop and change as the culture progresses.

Whereas traditional buildings can give a sense of the deepest roots of a culture, formal architecture is a representation of the grander aspects of the culture. Again going to ancient Egypt, we are only left with the more grand structures to reflect on the ancient culture. Furthermore, it could be argued that modern New York City structures do not reflect the true nature of the American culture. However, we can with some limited scholarly endeavor determine certain aspects of a culture by examining the built environment.

HOW DOES THE BUILT ENVIRONMENT REFLECT THE CULTURE?

When considering cultural architecture there are two predominant concepts: form and function. Although it is commonly believed that form follows function, careful examination with regard to the cultural landscape reveals that this may not always be the case. Function may not change across cultures; a dwelling is a dwelling in whatever culture is under discussion. Similarly with a place of worship or holy shrine, these do not change either. Because the function of a structure does not change between cultures, the defining cultural differences are seen in the form in which these buildings are constructed.

Dwellings

We will discuss a few examples of culturally determined structures starting with the humble dwelling. In this instance we take the term to refer to a place where a group takes shelter—from the elements and for protection. In western cultures we may refer to this as our home or house, but, as Oliver points out, "All houses are dwellings but not all dwellings are houses" (1987). To illustrate what he means consider the concept of a house—four solid walls and a door as the basic requirement. One can then go down a street, look at a building and say that is a house. However, in the Netherlands there are many windmills that have a dual purpose, a means of energy production and

Figure 3.6 Mongolian Yurt. (© 2011, Clouston. Used under license from Shutterstock, Inc.)

a place where the mill operator and family live. It would be wrong to call this a house. In other parts of the world diverse cultures used many and varied forms of structure as dwellings.

Culturally adapted dwellings can be seen in the tents and *Yurts* of the nomadic tribes of Asia and North Africa. These temporary dwellings reflect the nomadic culture of the peoples. They are easily dismantled and erected at a new site as the tribe follows their herds. "Although circular in design the yurt's interior reflects the social dictates of the culture group. The front half is strictly for those of low status while the rear is for the master and honored guests. To the left of the doorway is strictly for the ritually-pure, the males; while to the right is the impure—the females" (Humphrey, 1974).

Another form of adapted dwelling are those built some way above ground. Elevated dwellings are common in many parts of the tropics. These range from dwellings built on stilts a few feet above the ground to tree top structures tens of feet high. In coastal areas such stilt houses are built above the high water levels and reflect the cultural reliance on fishing. In other areas they are built to protect from the ravages of insects; while in south India the Urali peoples build their dwellings on stilts to keep their young women in seclusion. In Amazonian Peru, the communal dwellings of the Yagua people consist of little more than a wooden platform raised about five feet above ground, covered with a thatched roof and without walls or any form of interior partition.

Many cultures build their homes or dwelling places in the ground. Noble (2007) notes four broad categories of such dwellings: *pit houses*, which are partially below ground; *dugouts*, which are excavated in loosely compacted earth usually on sloping ground; *caves*, wholly underground excavations in relatively soft material, or *caves* in hard rock cavities that have been subsequently enlarged. An example of a cultural pit house is that of the Pawnee, a North American Indian tribe. Their lodges were excavated two to three feet and covered with branches and sod, with the entrance via a long, partly underground tunnel-like ramp. Dugouts though seem to be more of a temporary structure preferred by early settlement groups in North America, built into the side of a hill on newly acquired land prior to the building of more permanent structures.

Figure 3.7 Cave Dwelling. (© 2011, Connors Bros. Used under license from Shutterstock, Inc.)

When it comes to caves we tend to think of ancient cultures inhabiting such dwellings—the cavemen of Neolithic, Stone Age, times. Yet, in the late 20th and early 21st century there are still people who dwell in caves. Two types of cave are noted: those dug into soft rock and those in hard rock natural excavations, which have been subsequently enlarged. Soft rock cave dwellings are found in Tunisia, Turkey, and more widely in Northern China. The Chinese cave dwellings have been extensively researched and documented.

Cave dwelling in China has evolved over many thousands of years from Paleolithic times onward. Most prominent are the Löess cave dwellers of the Northern provinces. This area has become known as the cradle of Chinese civilization. It is in this region that Chinese Neolithic village culture and permanent agriculture appeared (Golany, 1992). The Löess formations are extremely thick in these parts and present a soft nature that is easily excavated. The dwellings are either dug into a cliff with an open terrace or are a series of rooms in the cliff around an open courtyard. The interior of the cave extends deep into the rock face with a horizontal depth of five-plus meters (18 feet). In cultural terms cave dwelling makes an ideal setting for the extended Chinese family, with its intimate ties and interrelationships. The design of the dwellings reflects this traditional form of living that fosters a sense of unity.

To many people, hard rock cave dwelling may be a thing of prehistory. Nevertheless, throughout the world there are still examples of people living in caves. Limestone is a particular favorite for cave dwelling. It is a hard rock but easily excavated. There are still many documented dwelling sites in the limestone areas of Europe, most notably in Spain and France. In the lower Dordogne valley of southwest France, one can see two and three room residences plus extensive communal networks. It is in this area that the cave dwelling dates back to prehistory with Neolithic paintings at Lascaux and the discoveries at Cro-Magnon.

The humble house may seem fairly ordinary to some in the west, but, throughout the world, dwellings built at ground level vary as the culture varies. Some may be small, one-room, circular

Figure 3.8   Sketch of a Russian Koshel. The living quarters are to the left and the farm buildings to the right. (M. Margaret Geib)

buildings constructed of mud and thatched with branches or grasses as found in rural areas of Africa and India. However, among the Hausa of Northern Nigeria where there is a large "extended" family, usually consisting of the male head of household who has several wives and numerous children. Each wife and her children will have a separate hut. The huts are grouped in a circular fashion around a communal cooking area and surrounded by a compound wall or fence. In this instance, the whole compound can be considered the dwelling.

In some cultures the dwelling is a long rectangular structure, usually built of wood. These structures can be for either single extended family occupancy or for larger groups or the village. Some of these large dwellings will have two entrances, one for men and the other for women. In other cultures, the dwelling may be internally divided according to social or family ranking. Here in these cultural adaptive dwellings, strangers rarely get more than a few feet into the dwelling while the higher ranked or socially intimated will get the furthest. This hierarchical separation exists in Chinese dwellings where seniority and age are the determinants, not gender.

In western cultures we mostly live in one or two story houses and like many other cultures throughout the world the living space is functionally and socially divided. There are spaces for living, preparing and eating food, bathing and sleeping. In Japanese culture all these functions are viewed as transitional. "Why possess three rooms when two are continually idle. Thus with light internal partitions and minimal furniture the functions of relaxing, dining and sleeping can be accommodated in a single room" (Earle, 1943).

The internal function of a dwelling may not always follow accepted norms, especially in agricultural communities where livestock are important. In fact, where more than function is required the form may be problematic. Take, as an example, **housebarns,** structures that house not only people but their animals, equipment, tools and food as well. Housebarns appear to be predominant among the cultures of Northern Europe from the Netherlands, where they are known as Gulfhaus, to the *Koshel* of Northern Russia.

Location, Location

Now that we have examined the form of the dwelling we can look at the fundamental aspect of the built environment location. When considering the location of a dwelling, many people will think first of the environmental impact. Is there a chance of flooding? Will there be enough sunlight? Is the site well drained? In many cultures the location of a dwelling is governed by specific cultural constraints. In some, the rules that govern site location are sacred while others relate to the social order. In India, for example, villages may be segregated according to caste or religion. In northwestern India, the houses of the highest caste are situated in the western part of the village, lower ranks on the east while the untouchables are segregated in the south. In many cultures, these rules have been incorporated into sacred documents or into regulations and procedures which are not sacred but are regarded by many as such—disobey or ignore at one's peril.

THE SPIRITUAL LANDSCAPE

Travelling through the Old World of Europe or Asia, one can observe the remnants of past cultures mostly in the spiritually sanctified structures they have left. Every culture, past and present, has or had sacred structures or buildings. Some past cultures have left an indelible mark on the cultural landscape. Modern Cultures are also apt to erect or modify structures for some holy purpose. We will consider the beliefs and religions later in the text but at the moment we will concentrate on the structures that relate to past and present beliefs. Many of these sites are either destinations for pilgrimages or the objective of tourists.

The New Stone age peoples of Britain erected **henge monuments,** such as Stonehenge and the great stone circle at Avebury and buried their noble dead in long, above ground, tombs called barrows. The Egyptians built many buildings of a spiritual nature including temples and the Pyramids, which are not places of worship, but the spiritual resting place of the pharaohs. The Greek built sacred sites to worship their deities as did the Romans.

The Mayans of Mesoamerica built pyramids which were active sites of religious activity where rituals happened, possibly daily. These Mayan pyramids are unlike their Egyptian counterparts which are strictly grave markers. Also in Mesoamerica are the ruins of former cultures notably at Machu Picchu, a spiritual site that predates the Inca culture.

Figure 3.9   Stonehenge. (© 2011, stocker1970. Used under license from Shutterstock, Inc.)

In Asia there are many spiritual sites that can be traced back to early Buddhism. Scattered across the Indian subcontinent are several stupas, built of stones or bricks to commemorate important events or mark important places associated with Buddhism. They were built 2,000–3,000 years BCE. In Northern China's Shanxi province there are many Buddhist relics; grottos, statues, rock carvings and temples—notably the Hanging Temple at Jinlonkou built 1,500 years ago. In Cambodia the spectacular Angkor Wat temples are a tribute to the culture that built them.

In the Christian era churches are time capsules showing cultural development through the ages, reflecting the architecture, art, wealth and politics of the society who built them. Many churches are built on an east-west axis in the shape of a cross, with the chancel and sanctuary (front section from where the service is conducted) and nave (where the congregation gathers) crossed by the transept ("arms" of the cross). The most important church in Christendom is St. Peter's Cathedral in the Vatican City. Smaller parish churches dot the countryside in most Christian cultures.

In Judaism the *synagogue* is more or less the equivalent of a church. In actual fact it is more than the Christian church. It is the center of the Jewish religious community: a place of prayer, study and education, social and charitable work, as well as a social center. There is no set blueprint for synagogue design; the architectural shapes as well as interior designs vary greatly. In fact, the influence from other local religious buildings can often be seen in synagogue arches, domes and towers. The styles of the earliest synagogues resembled the temples of other sects of the eastern Roman Empire. The surviving synagogues of medieval Spain are embellished with Moorish art, while some of those in Eastern Europe have a Gothic appearance.

Mosques are impressive examples of cultural architecture. With their soaring minarets, vast domes and decorations, they are examples of cultural art and architecture. The Mosque is the center of communal life and is usually the center of the neighborhood or urban area. Very few mosques do not have shops and commercial activities in the streets around them, with people's homes often lying in a circle outside the mosque and the shops. It is rare for mosques to lie in open areas. Apart

Figure 3.10   English Parish Church. (© 2011, Leonard Peacefull)

Figure 3.11   One of South India's great temples is the celebrated Meenakshi Temple, at Madurai. Situated on the banks of river Vaigai, Madurai has a rich cultural heritage passed on from the great Tamil era more than 2,500 years ago.

(© 2011, Arvind Balaraman. Used under license from Shutterstock, Inc.)

from being a place for prayer, other social functions are connected to mosques: schools, law courts, hospitals and lodging for travelers. This pattern is based upon the mosque in Madina.

Of all the spiritual buildings, those which are the most ornate and thus portray variations in the development of a culture are the *mandir* or *temples* of the Hindu. Stonehenge, an English parish church, or a Buddhist Stupa are bland compared to the decorations found on many mandir. The earliest temples dating back to the 5th century BCE were possibly made of clay with thatched roofs made of straw or leaves. Stones or bricks came much later. In Hinduism, unlike other main religions, it is not mandatory for a person to visit a temple since all Hindu homes usually have a small shrine for daily prayers. The temple serves as a sacred meeting place for the community to congregate and revitalize their spiritual energies. Large temples are usually built at picturesque places, especially on river banks, on top of hills and on the sea shore. Smaller temples or open-air shrines can be found just about anywhere—by the roadside or even under a tree. Many mandir of southern India have Gopurams—towers, exquisitely decorated with painted sculptures and carvings with a variety of themes derived from the Hindu mythology, particularly those associated with the presiding deity of the temple.

OTHER ADAPTATIONS OF THE CULTURAL LANDSCAPE

There are many other culturally reflective structures scattered across the globe. These include defensive works such as castles and fortifications like the Great Wall of China; agricultural buildings—barns, hayricks, sugar bushes and outhouses; leisure facilities like the Colosseum in Rome or the Roman bath house in Bath, England, even modern facilities as in the Olympic stadium in Beijing. The Romans left many cultural impressions on the landscape throughout Europe that are

Figure 3.12   Warwick Castle. (© 2011, Leonard Peacefull)

Figure 3.13   Pont du Gard. (© 2011, Leonard Peacefull)

Figure 3.14   Disused Canal and China Factory—Coalport, England; Now a Living Museum. (© 2011, Leonard Peacefull)

still extant today. Many of Europe's straight roads follow the line of the Roman roads and in some places the original road takes off across the country away from the modern highway. In the south of France is a colossal feat of engineering, the aqueduct of Pont du Gard, that carried water from the Alps to the dry part of Languedoc.

In the post industrial revolution era, there are many buildings that reflect the culture, the growth of manufacturing and wealth. Old mills and factories are still scattered across the landscape of Europe and North America, particularly in New England. Canals, with their associated locks and railways, many now disused, have left their mark on the landscape. It is a sign of the ability of cultures to adapt that many of these buildings now have a different use. In certain parts of the developed world, warehouses are now dwellings. A similar fate has befallen several disused churches in Britain, while railway stations have been adapted to other uses, such as restaurants, lodgings and even museums such as the Musee d'Orsay in Paris.

LANDSCAPE AND CULTURAL DIVERSITY

Understanding the cultural landscape is an integral part of the process of examining and appreciating cultural diversity. Cultural landscapes have been created by a diverse range of cultural groups. Places of contemporary social significance are part and parcel of how indigenous people construct their social identities today, but are also intimately linked to longer term patterns of maintaining social identity.

A connection with familiar landscapes forms a culture's identity, as people feel they *belong* to one place, one region, one country. Thus, a cultural landscape is more than just the sum of its physical places; it is equally concerned with the environmental space between places and the meaning given to them through history. Additionally, the historical evidence found in the documentary and oral histories are woven into each culture. The deeply social nature of relationships to place has always mediated people's understandings of their environment and their movements within it and is a process which continues to inform the construction of people's social identity today. Landscape values accrue historically.

In addition to traditional cultural development, we must consider the repercussions brought about by contact with outsiders, who placed very different cultural values on the same physical landscape. The expansion of the colonial frontier radically altered the ways in which indigenous people could relate to the old physical places and spaces and brought new layers of meaning to the same places. For example, European colonists were engaged in the process of actively constructing a new world for themselves, in many instances in complete disregard for the indigenous peoples. It is the dynamics of these competing social landscapes, perhaps more than any other, which have created the potentially conflicting heritage values of today.

In the next chapters we broaden the theme of the cultural landscape by looking at the changes that occurred with the development of agriculture and the cultural impact these changes brought about. We will also examine the evolution of many of the traits which are now taken for granted along with the ascendance of the urban way of life so familiar to everyone today.

UNESCO World Heritage Cultural Landscape Categories (UNESCO 2008)

Three main categories of cultural landscapes have been identified by UNESCO for World Heritage Landscapes. The criteria for each are:

1. Clearly defined landscape designed and created intentionally by man. The most easily identifiable, this category embraces garden and parkland landscapes constructed for aesthetic reasons which are often (but not always) associated with religious or other monumental buildings and ensembles.
2. Organically evolved landscape. This results from an initial social, economic, administrative and/or religious imperative and has developed its present form by association with and in response to its natural environment. Such landscapes reflect that process of evolution in their form and component features.

    They fall into two sub-categories:
    a. a relict (or fossil) landscape is one in which an evolutionary process came to an end at some time in the past, either abruptly or over a period. Its significant distinguishing features are, however, still visible in material form.
    b. continuing landscape is one which retains an active social role in contemporary society closely associated with the traditional way of life and in which the evolutionary process is still in progress. At the same time it exhibits significant material evidence of its evolution over time.

3. Associative cultural landscape. The inclusion of such landscapes on the World Heritage List is justifiable by virtue of the powerful religious, artistic or cultural associations of the natural element rather than material cultural evidence, which may be insignificant or even absent.

CULTURAL LANDSCAPE EXERCISE

Look at this picture and identify 10 aspects of the cultural landscape.

1 _____

2 _____

3 _____

4 _____

5 _____

6 _____

7 _____

8 _____

9 _____

10 _____

KEY TERMS

Natural Environment	Built Environment	Cultural Landscape
Material Culture	Vernacular Landscape	Ethnic Architecture
Formal Architecture	Vernacular Architecture	Dwelling
Production Mode	Communication mode	Built Mode
Material Mode	Housebarn	Henge Monument

DISCUSSION TOPIC

Preservation of the cultural landscape encourages succeeding generations to consider their cultural heritage.

CHAPTER 4

The Cultural Development of Agriculture

Geographers and Anthropologists divide human culture into four distinct developmental stages: the food-gathering and hunting cultures; herding cultures; agricultural cultures; and urban cultures. With each stage comes an increasing complexity of material goods, social organization, higher population densities and a greater interference with the natural environment. Consequently, if we accept that the first two are fairly primitive stages of cultural development we can ascertain that most cultures develop from the sedentary activities associated with agrarian societies. Agricultural, then, is the basis of all cultural development. In fact, we can trace most of our many cultural traits; writing, fabrication of artifacts—manufacturing, settlement patterns—urbanization, sociofacts—common laws and rules of societies—back to the development of agriculture.

In the last chapter we looked at the impacts of cultures on the landscape in the form of buildings or material culture. However, careful examination of the landscape will reveal that cultures have also left their mark in the land itself. These marks are in the patterns of agricultural activity of former cultures that can still be seen in many parts of the world. In the Negev desert in present day Israel, Rubin (1991) reports the remains of farmhouses, agricultural field systems, cisterns and wells. In parts of Central America, Mayan canals are thought to represent an intensive hydraulic system to extend the cropping period into the dry season (Jacob, 1995). In Britain, the **ridge and furrow** system of the Saxon three-field system can still be seen in aerial photographs and in close observation on the ground (Mead, 1954). Thus agriculture has left its mark on the landscape as well as on the development of humankind.

THE ORIGIN AND DIFFUSION OF AGRICULTURE

The beginning of agriculture—the domestication of plants and animals—is one of the earliest major cultural innovations to effect human development. This is the third stage in the development of cultures after hunter-gatherers and nomadic herding. At the end of the last Ice Age, the population slowly increased. Climatic fluctuations adversely affected established plant and animal food sources. Consequently, people, in more than one world area, experimented with the domestication of plants and animals. There is no agreement on whether the domestication of animals preceded or followed that of plants, the sequence may well vary in different areas. What appears certain is that animal domestication began during the middle part of the Stone Age (the **Mesolithic** era). This was not as a conscious economic effort by humans but a development from the keeping of small or young wild animals as pets plus the attraction of scavenger animals to the refuse of human settlements. The docility of some animals meant they were easily herded—a course of events that ultimately led to full domestication.

This **domestication** process gained momentum in the latter **Neolithic** (The New Stone Age) period. The domestication of pigs and goats probably occurred in southeastern Turkey and the Near East about 8000 BCE; sheep were also first domesticated in Turkey (c. 7500 BCE), with cattle in Greece and the Near East later. Although there is evidence that animal domestication diffused from limited source regions, once its advantages were learned, additional domestications followed, aided by the large variety of species able to be domesticated. Not all domestication happened at the same time; for instance, distinctive species of cattle were domesticated in India, north-central Eurasia, Southeast Asia and Africa at different times, while several varieties of pig and wild fowl were also domesticated in other places at varying times.

The **domestication of plants,** like that of animals, appears to have occurred independently in more than one world region. It is accepted that food crops were first cultivated in the Near East 12,000 years ago and dispersed rapidly from there across the midlatitudes. There is also evidence that people in Africa were raising crops of wheat, barley, dates, lentils and chickpeas about the same time. In other parts of the world farming is a more recent occurrence, appearing in Mexico about 5,000 years ago.

The development of agriculture was most likely the logical extension of plant selection for nonfood activities. Poisons obtained from plants and applied to hunting arrows made food gathering easier and more certain. Plant dyes and pigments were collected and prepared for personal adornment or decoration of artifacts. Medicinal and mood altering plants and derivatives were cultivated by all early cultures, while evidence exists that early cultivation of grains was not for grinding and baking as bread, but for brewing as beer for religious and nutritional reasons. It was probably healthier to drink beer or ale than putrid water. These factors may well have been the threshold for sedentary agricultural activities. The destruction of wild game habitats favored the selection and cultivation of short-season annual grains and legumes whose seeds could then be stored and planted during cooler, wetter winter growing seasons.

In these early societies, women were assigned the primary food-gathering role. They were the first to become familiar with nutritive plants. Consequently, their role in initiating crop production to replace less reliable food gathering is certain. It is safe to say that women were major contributing innovators of technology in food preparation and clothes production as well as being inventors of such useful items as baskets, jugs and pots, baby slings and yokes for carrying burdens, such as water pails.

Agricultural Hearths

From the above we can conclude that there are several areas that would qualify as agricultural hearths. While archaeological evidence mostly points to the fertile crescent of the Middle East as the probable source region, Carl Sauer (1969) offers five criteria for the establishment of hearth areas that are somewhat contrary to archaeological evidence. In the first, he claims domestication could not take place in areas of food shortages; because domestication implies experimentation, the experimenters could wait on the results; therefore, an abundance of food was necessary. Following on, his second criterion stipulates that agricultural hearths were in areas with a great variety of plants and animals because there had to exist a large enough gene pool for experimentation and hybridization to occur. These first two examples are easy to accept, however, his next criterion bears little conviction. He suggests large river valleys are unlikely hearth areas because their settlement and cultivation require rather advanced techniques of water control. This is not a convincing argu-

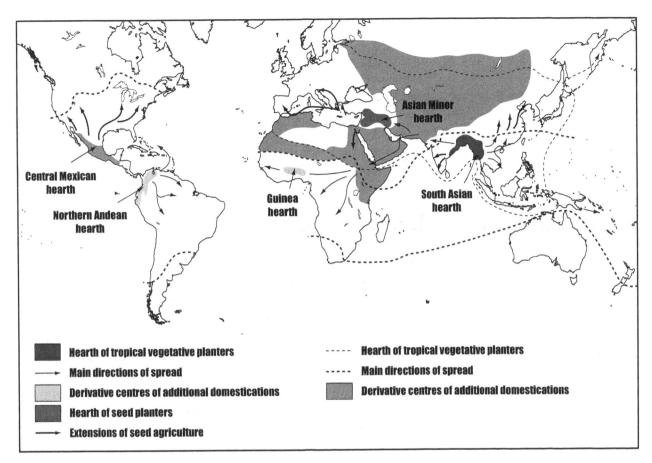

Figure 4.1 Agricultural Hearths after C. O. Sauer.

ment as evidence shows that early farmers were active in the Nile and Euphrates valleys long before any other areas were cultivated. Therefore, he stipulates that hearths must be restricted to woodland areas where spaces can readily be cleared by cutting and burning trees. The thinner soil in these places would be easier to till than the grasslands where the sod was probably too tough for primitive cultivators. There is now expert opinion that these areas were probably on the edge of woodlands rather than deep in to the interior. Furthermore, this argues against the notion of the **Fertile Crescent** as a hearth since there was probably light scrub and bush that was easily removed. His final condition, that the original cultivators had to be sedentary, is fairly obvious as nomadic groups would not meet the conditions for cultivators, nor, probably, did the areas they inhabited. Although there is sometimes hostile response from archaeologists to his view, Sauer's main argument is that we need to think about the geographical conditions under which cultivation and domestication might occur. The landscape had to be right; there had to be a suitable climate to induce agricultural diversity and most likely rivers to provide a regular supply of fish for a sedentary settlement during the "off-seasons."

The Diffusion of Agriculture

Although in the previous section we have highlighted different opinions on the early development of agriculture, what we can conclude from archaeological evidence is that agricultural communities spread from the hearth areas of the Middle East in three directions. First, there is the spread into the steppes of Eurasia where tillage gave way to pastoralism and in some cases true nomadism. Evidence for this can be seen with the Mongol tribes of Central Asia who still wander the steppes of central Asia to this day. Second, we see a westward movement distinguished by the Celtic, Germanic and Slavic peoples who took rye and oats into the northern climates. It is in northern Europe that animal husbandry developed; the intense manuring of the ground encouraged the development of cultivation. Lastly, we note the development along the shores of the Mediterranean where wheat and barley became the staple grains. In terms of animal husbandry, sheep and goats were of greater importance than cattle and horses. Where the land became more ragged and unsuitable for cropping, cultures turned to herding, a practice still common in Mediterranean lands today.

THE DIVERSITY OF AGRICULTURE IN THE MODERN WORLD

From the humble beginnings of seed planting and domestication, agriculture has developed in different formats according to the cultural development of its practitioners. Leaving aside the hunter-gatherers, of whom there are few groups surviving in the modern world: the Bushmen of the Kalahari Desert in Namibia, the Aboriginal tribes of Australia and some culture groups in the Amazon basin, we recognize two basic classifications of agricultural activity based on fundamental economics. These two groups are subsistence agriculture, where the practitioners are barely or just supporting themselves and commercial agriculture where the farmer is able to produce a surplus and make a profit. The latter situation can range from the small family farm that supplies a local market to the huge commercial enterprises that supply food processing plants of various kinds. For the purpose of this section we have subdivided these classifications into groups.

Subsistence Agriculture

There are many parts of the world where cultural development has not progressed at the same rate as in the developed countries of the "west." There are several factors that impact this lack of development. Remoteness: Peoples far from the main areas of population have not had a reason to develop trade patterns and consequently are content to continue as always. Climate: Areas of poor climate, excessive drought, too little sun, or too much sun can affect the growing season and therefore the crop yield. The customs and beliefs of a particular culture may prevent them from interacting with the outside world. These factors have led to people just surviving or subsisting on what they can produce. We can recognize three forms of subsistence agriculture; shifting cultivation, pastoral nomadism and intensive subsistence.

Shifting cultivation, often called slash and burn, was probably the original form of cultivation dating back 10,000 years. It is a practice where a group from the tribe goes out and cuts down the forest, burning the brush so that the nutrients released from the burning enrich the soil. The main group then comes along and cultivates this patch which may be only one or two acres. However, successful cultivation will only last at most five years as the continued planting will deplete the nutrients in the soil and the crop yield will suffer. Then another area of the forest has to be cleared and burnt and the group cultivates the new patch. Original outside observers of this form of cultivation regarded it as indiscriminate destruction of the forest. However, prolonged research has shown that after a period of time, around 20 to 25 years, the tribe returns to the first patch and so the cycle continues. It is recognized that these cultures practice a very primitive form of rotation agriculture and therefore rather than destroying the forest they are preserving it.

Nomadic Pastoralism is a form of subsistence agriculture practiced in areas of the world where the climate or the terrain is not conducive to cultivation. This practice takes the group, which may be a family or small tribe, from one watering hole or grazing plot to another. It is estimated that only about 15,000,000 people throughout the world rely on herding in this way; however, the practice may cover about 20 percent of the surface, a larger area that all the cultivated land. Culturally, these peoples look to their animals as important to their wealth. A Masai herdsman counts on the number of cows he owns to give him status; a Masai woman looking for a husband would consider one with the larger herd as preferable. The nomadic herdsmen rely on their herds, camels, goats, sheep or cattle to supply them with milk, wool for trade and their hides for shelter. Only when an animal dies do the herdsmen take the hide or use the meat. So where do they obtain their food? Most nomadic tribe live on grain that is either grown by a section of the tribe in a semi-nomadic lifestyle whereby the women and children remain in an area where there is cultivatable land and water while the men move off to find fresh pasture. This lifestyle is more common among nomadic tribes especially in East Africa and the southern Sahara regions. True nomads would purchase grain with animal products, wool, hides, etc.

Intensive Farming usually refers to obtaining the most from a very small space; it usually requires a large input of labor especially in areas of Southeast and East Asia where a large rural population relies on agriculture. **Intensive subsistence** farming is practiced where cultivatable land is at a premium, where large families must produce enough food for their survival and is therefore located in the more heavily populated parts of the world. Intensive agriculture usually employs large draft animals to pull ploughs, intensive manuring and fertilizers and water management. One example usually associated with this form of agriculture is wet rice cultivation.

Wet rice cultivation is very labor intensive and is usually carried out by the whole family. We regard this as a subsistence form of agriculture because over half the world's rice crop is consumed by people living off the land on which it is farmed. It is carried out in the highly dense rural areas of Asia where farms are small and fields are scattered. The fields are normally referred to as a rice padi although in some parts of Indonesia these fields are called Sawahs. A characteristic of the padi is that it is generally a small irrigated plot. In some areas, notably southeast China and Japan, terracing of the landscape predominates. Terracing is a method of shaping the hillside to capture water and to stop soil erosion. It takes the form of a series of level benches or large steps cut into the hillside with large retaining walls in between. In parts of southeast China and Taiwan, double cropping—growing two crops in the same field in the same season—is possible due to the favorable climate. Elsewhere in south Asia multiple crops can be grown as one summer and one winter crop. The cultural importance of this crop is significant in areas of its cultivation. Although it is labor intensive, it has a high yield per hectare with a high calorific content, the crop can feed significantly large numbers.

Commercial Farming

When a farmer produces a surplus of a crop that he is able to sell, he then enters into the realm of the commercial farmer and the farm becomes a business. Commercial farms can range from small family enterprises to large agribusinesses. When a farmer begins to produce a surplus he then moves into a different phase of cultural development. He becomes self-sufficient.

Mixed farming occupies most of the mid-latitudes in Eurasia, in North America east of the 98th line of longitude, South America, South Africa and Australasia. Modern **mixed farms,** which range from **small holdings** to large commercial enterprises, share one thing in common; they raise their crops and livestock for sale to the market. But they are also major players in buying inputs such as fertilizers and feed for the production of these goods. In most of Western Europe and North America, mechanization of the many processes has led to a decline in the number of people employed in this activity; whereas in many parts of Eastern Europe farmers still cling to the culturally imposed system of the communist era, high labor amounts and little mechanization. France has a particular version of mixed farming called **polyculture** that is characterized by a wide variety of crops integrated with livestock rearing. This practice remains a fundamental trait of agriculture over wide areas of the country, where many of the farms are divided into small parcels of land. Within these small farms there is grown a variety of crops accompanied by a miscellany of dairy animals, pigs and various fowl. For instance, in a small 10 hectare plot in The Massif Central a variety of crops were observed: wheat, sunflowers, corn, a vineyard and a walnut grove. The aim of this method of farming and the variety of produce, was to transform the rural economy and to afford security against individual crop failure. The establishment of polyculture also represented an intensification of production which permitted a form of commercialization and gave the peasant farmer individuality and a sense of security.

In the Mediterranean basin and in other parts of the world that experience a similar climate, a form of mixed farming know as Mediterranean is practiced. Here, farmers have to contend with poor soils and dry summer climates. Grain is the chief crop in the western end of the region, while other crops such as olives and grapes predominate on the less favorable land. Livestock, predominantly sheep and goats with fewer beef and dairy cattle, occupies a large proportion of the farmland although the numbers are small as a result of the carrying capacity of the land and the dry summer climate. Transhumance—the movement of livestock from a winter pasture in coastal areas to an upland summer pasture—occurs in many parts of the region and although this practice is becoming less important it still plays a part in the culture of the people of this region.

Away from the areas of mixed farming, the agricultural activities tend to be monocultural, relying on one specialized farming practice. These activities are extensive in nature—large agricultural enterprises, usually agri-business, run by a corporation. In areas where the soils and climate allows, grain crops such as wheat are produced on large prairie style farms while in drier less flat areas ranching predominates.

The most extensive single crop grown around the world is wheat. Wheat has many favorable properties apart from being a staple food source for breads, cakes and pasta. Wheat can be stored for considerable periods of time without spoiling and can therefore be transported over longer distances to remote markets. Because of its storability and variety of uses, wheat has always commanded a higher market price than other cereals such as barley, rye and oats. In North America, Argentina, Australia, the Ukraine and Russia, wheat is grown on a large scale. For example, farms in North America average around 500 hectares (about 1,200 acres), while those in Russia can be up to 4,000 hectares. Yields per hectare in these areas, ranging from 1.6 to 3.0 tons/hectare, are quite low compared to Western Europe where yields are as much as 7.5 to 8.0 tons/hectare. The reason for this difference is that most large-scale grain farms are in areas of low variable rainfall.

Extensive commercial grazing otherwise known as **ranching** seems at first glance to be a more economically advanced form of pastoral nomadism (Grigg, 1974). Ranching is mostly confined to the drier lowlands of the World where the carrying capacity of the land is relatively low; the high plains of the western United States, the llanos of Venezuela, the Karoo of South Africa and the interior of Australia. The Canterbury Plain and the high country of North Island in New Zealand, which experience more rainfall and a higher carrying capacity, have many large farms. Beef and

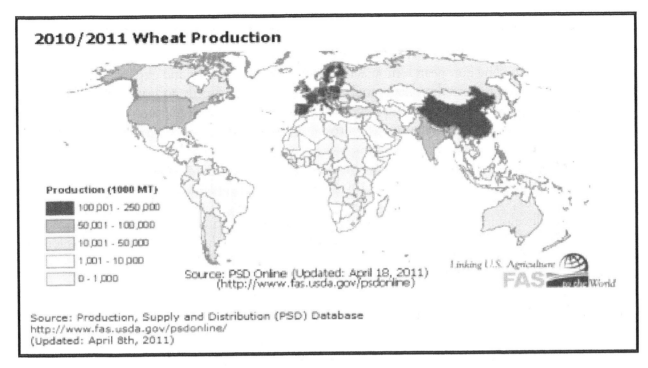

Figure 4.2 World Wheat Production. (USDA, 2011)

sheep are the two animals most commonly found on ranches. The main reason for the concentration on these two animals is both cultural and economic; there was a demand for their meat and wool in major urban areas of the United States and Europe, particularly in the United Kingdom. Developments in transportation, such as refrigerated box cars and ships, greatly assisted the growth and demand for ranch products. With a high demand for their product and a low-carrying capacity, ranching is therefore a large-scale operation. In Australia, for instance, a sheep station can average 8,000 hectares while a cattle ranch can be over 10,000 hectares. Large expansive properties are required because of the nature of the available forage in some parts. It may take 6–10 hectares to support one head of cattle, while one sheep may need in excess of 3 hectares. These figures can range higher or lower depending on the nature of the forage area.

Plantation agriculture has also developed out of cultural/economic necessity. The large populations are concentrated in the more temperate latitudes while crops such as rubber, palm-oil, coconuts, cacao (from which we obtain cocoa) and coffee are confined to the tropical latitudes. We may also refer to these crops as industrial crops because they all find their way into a manufacturing process of one form or another. The large industrial concentrations of Western Europe and North America needed these crops to sustain their economies. Consequently, many plantations are owned by large multinational corporations such as Bridgestone, who have rubber plantations in Indonesia and Nestlé, who operate large cacao plantations in Cote d'Ivorie. Besides the corporate plantations, a large number are of smaller holdings owned by a single family. Other plantation crops include tea grown in more temperate climates especially in the foothills of the Himalayas. This plant was introduced into Assam, India, by the British East India Company to satisfy a growing cultural demand in Britain and Europe. Originally a drink of the aristocracy, by the beginning of the 19th century it had filtered down to the lower classes and had become the hot beverage of choice.

Agriculture Today

Between 2005 and 2008 there was annual growth of around 10 percent in the value of world trade in agricultural goods. The European Union is the largest global importer and exporter of agricultural goods and particularly imported products such as tropical fruits, oilseeds and oils, fruit and vegetables. While the United States is a major exporter of food crops such as maize, soybean and wheat, China, on the other hand, is an importer of many agricultural crops such as soybeans, palm oil, natural rubber and cotton.

Many of the developed world countries have encouraged the liberalization of trade for agricultural products with the developing world in the hope this would contribute to sustainable economic growth for both exporting and importing countries. Recognizing the crucial role that agriculture plays in many developing countries, the developed world has granted extensive market access to agricultural imports from the developing world countries, which includes free market access for produce from the world's poorest countries.

The Cultural Impact of Agricultural Development

Wherever the first agricultural communities were located, the impact of permanent agriculture on the organization and density of the human population is evident. A formal agricultural system increased the reliability and volume of the food supply enabling a given area to support more people. Cultures no longer had a need to concentrate on food production but could branch out to non-agricultural activities. As food surpluses became available, goods began to be exchanged; pottery, weaving, jewelry and weapons were bartered and traded over long distances.

Having created a food-producing rather than a foraging society, cultures undertook the purposeful restructuring of their environment. They began to experiment with plant modification and breeding of animal species. They learned to manage the soil, the terrain, water and mineral resources and to use animals to replace human energy. Humans learned the arts of spinning and weaving plant and animal fibers. They learned to use the potter's wheel and to fire clay and make utensils. They developed techniques of brick making, mortaring and construction and they discovered the skills of mining, smelting and casting metals. They used metal to make tools and weapons; first of copper and later the alloy of tin and copper that produced the harder, more durable bronze.

As time progressed, cultures began to alter at an accelerating pace and change itself became a way of life. This cultural change did not occur at the same time around the world for all peoples. With these technological advances a more formal economy emerged that resulted in the foundation of more complex structured cultures; people began to live in towns and cities.

KEY TERMS

Neolithic

Nomadic Pastoralism

Intensive Farming

Subsistence

Ranching

Mixed Farming

Mesolithic

Fertile Crescent

Shifting cultivation

Commercial Agriculture

Mediterranean Agriculture

Small Holdings

Domestication

Plantation Agriculture

Ridge and Furrow

Polyculture

QUESTIONS

1. What are the significant implications of religious influences on agricultural activities?
2. How are various cultures reflected in modern agricultural practices?

DISCUSSION TOPIC

The lesser developed countries of the world are dependent on the markets of the developed world, forsaking their own internal markets, which in certain areas has led to starvation within the local community. Should these lesser developed countries be so dependent on the developed world markets that their own people starve?

CHAPTER 5

Urbanization

Ronald Runeric

In the previous chapter, we discussed the cultural changes that occurred with the advances of agriculture. These developments also brought about changes in the way societies were organized. The cultural forces that had previously scattered small groups of people over large areas were no longer prevalent; isolation gave way to a degree of agglomeration, people began to settle in agricultural villages. Archaeological evidence has shown that around 8000 BCE hill-farm communities of northern Mesopotamia had densities approximately 70 people/km² (180 people/square mile).

Movement by people into organized communities led to an increase in population density in some areas. This was several hundred times greater than the pre-agricultural communities. By 4000 BCE the total population of the globe had probably reached around 90 million. The greater part of this population was most likely concentrated in areas where village agriculture was intermingling with and slowly replacing food-gathering and hunting. Such areas certainly included a belt stretching from Western Europe and the Mediterranean through the Middle East to western India, northern China, Indonesia and Central America. Outside this area population changed little from its pre-agricultural pattern.

In the villages people acquired new skills. They learned to spin and weave plant and animal fibers and to make pots and other utensils. They developed skills in construction mining, smelting and casting metals. Thus, occupations became increasingly specialized; weavers and clothes makers, potters, builders, metalworkers, merchants and scribes and in some areas warriors were needed to protect the society. With such technical advancements, a more complex culture appeared and a more formal economy emerged. A stratified society based on specific roles replaced the inequality of hunter-gathering economies.

As people gathered together in larger communities, new and more formalized rules of conduct and control emerged, especially important where the use of land was involved. We saw the beginnings of social structures, governments and laws. The protection of private property became important as a sedentary life generated more property than that of the nomad; as did the enforcement of the rules of societies increasingly stratified by social privileges and economic status. The social and technical revolutions that began in the Neolithic period were initially spatially confined. The new technologies, the new ways of life and the new social structures diffused from those points of origin and were selectively adopted by people who came into contact with them.

THE DIFFUSION OF URBANIZATION

Early cities developed because agricultural surpluses allowed them to grow and occupations in addition to farming were required for people to live and thrive. The work of soldiers, leaders, weavers, merchants and builders allowed more people to survive and live more comfortably than they would live on farms of the day. Modern cities have evolved on these ancient models; cities are still concentrations of human population and economic activities in a central place. Centrality does not necessarily mean that a city is located exactly in the middle of its hinterland, but that it is universally accessible to the entire region. Centrality allows urban populations and commerce to draw upon the resources of surrounding locations. Cities thrive based on the extent of their export functions (Jacobs, 1984). In this context, "exports" are goods and services sold to people outside a given city. This concentrates wealth within the city, so it may become considerably more prosperous than its hinterland. Urban import and export functions depend on the two spatial dimensions known as "site" and "situation." Site refers to the physical, social and economic conditions at a given location; situation is the combination of location-related assets of a given place relative to all other places with which its population interacts, e.g., proximity, good highway connections, etc. Although advantages of site and situation are often studied separately, they operate in a complimentary manner to help the development of some cities and hinder others.

Site advantages strongly influence the establishment, growth and importance of cities. For ancient cities, easily defended sites were vital for a city to survive and thrive; economic site advantages may be comparably important for the survival and vitality of modern cities. "Break-of-bulk"

Figure 5.1   A massive container ship entering a port. (© 2011, Jurand. Used under license from Shutterstock, Inc.)

points comprise a major category of physical benefits that some sites possess over others. A break-of-bulk point is any location where commercial products must be transferred from one mode of transportation to another. Ports are break-of-bulk points where goods are transferred from ships to land transportation. Many modern examples of the importance of high-quality harbor locations are obvious, for example, New York, London, San Francisco, Los Angeles, Sydney. Social conditions can provide conditions for cities to grow and thrive. Availability of large supplies of cheap labor attracts labor-intensive industries (industries that require large numbers of workers) to cities in less developed countries (LDCs). On the other hand, workers with specialized education and skills are most abundant in urban areas of more developed countries (MDCs).

Urban Areas Today

In developed countries, the urban-rural proportions of populations are generally stable. The percentage of total population living in urban areas in the countries of the European Union combined is about 77 percent, in the United States 82 percent and in Canada 81 percent. In both North American countries, slightly more than 75 percent of their populations lived in cities by 1970. The major impetus for rural-to-urban migration was industrialization, which provided employment at the same time that agriculture became more mechanized and less labor-intensive than during the 19th and early 20th centuries. Like the first ancient cities, growth of cities in LDCs is closely linked with the level of prosperity of agriculture in the hinterlands. As long as agriculture provides adequate livings for rural people, they continue farming, but when food supply increases, farm prices decline, causing rural-to-urban migration. Population growth in LDCs is also a push factor in this process; as land becomes less available, rural people are forced to migrate to cities. The Industrial Revolution continues in LDCs and the number of middle-income people is slowly rising in rapidly-growing countries such as India and China. However, these places are home to large populations of impoverished unemployed people or workers in marginal occupations. In those situations, cities grow in population only and the quality of urban life may decline.

Natural site characteristics influence cities' attractiveness for migrants by affecting economic development. Economic opportunities in large cities in Southeast Asia attract migrants from rural areas. In addition, semi-skilled workers migrate across national boundaries from low-wage to high-wage countries, such as Vietnam, Philippines and Myanmar to Brunei, Malaysia, Thailand and Singapore. Consequently, the larger the level of immigration, the more acute is the need for housing, sewage and refuse disposal and transportation systems.

Site characteristics influence the health of urban residents and thus the vitality and even survival of cities. According to Vearey et al. (2010): ". . . the health of urban residents is more than the risk factors of individuals. . . . Both the social and physical environment of cities, combined with health and social service systems. . . . form the primary determinants of the health of urban populations." Housing availability is an important attribute of any urban site. Housing problems for the working poor are abundant in MDCs and even more pervasive in LDCs. Location matters: the quality of sanitation and urban services varies among and within cities and the health of urban residents can depend on their locations. Dilapidated inner-city housing occupied by low-income people is visibly abundant even in American cities, such as Philadelphia, while upper-income residents live in suburbs and relatively distant exurbs. In both MDCs and LDCs, living adjacent to health hazards is a condition related to large disparities of personal income. Although the number of middle-income people is growing in some countries such as China and India, there is still a large number of working poor.

In LDCs, housing can be classified as either formal or informal. "Formal" housing generally refers to buildings intentionally constructed for residential occupation, which comply with laws regarding safety and sanitation. Formal housing is generally occupied by middle- and upper-income residents. Formal housing can be modern, high-quality, or even opulent for the growing number of relatively wealthy people living in metropolitan areas of LDCs. Many Southeast Asian "edge cities"—small cities located on urban fringes, similar to suburban towns in the United States—have grown much faster than established large cities they border. Between 1950 and 2000, the population of Jakarta, Indonesia, grew at a rate of 4 percent, while populations of some fringe areas of the city grew at 18 percent. These edge cities include upper-middle income urban amenities, similar to the United States that includes golf courses and shopping malls. Yet, slums are extensive and house a large proportion of urban residents in most LDCs.

Informal housing generally includes residences on land which the occupants neither own nor rent; it can also be a temporary or makeshift shelter erected on rented land. In order to settle land without legal authority, vacant land must be identified, either by potential settlers or by an entrepreneur who may construct makeshift housing for rent. The land would be either owned by a municipality or by an individual absentee owner. In order to be vacant, the land would have no obvious utility; it could be in a floodplain or other area on which building would be inadvisable or illegal. Some informal housing is constructed with the consent of the landowner. The land itself is rented and the renter constructs a makeshift shack; of course, slums can be created by constructing many such shacks on one or more plots of land. Informal housing also refers to disused

Figure 5.2   Favela in Manaus, Brazil. (© 2011, guentermanaus. Used under license from Shutterstock, Inc.)

buildings—e.g., warehouses, factories, stores, etc.—occupied as residences either with or without landowners' consent.

The urban fringes in Latin America are not locations for opulent suburbs as in North America; cities there are often ringed by slums, known as "favelas" in Brazil. Favelas are squatter settlements that are allowed to develop where land is unsuited for profitable uses. In addition, use of abandoned inner city buildings for informal housing is growing in Brazil. Occupants sometimes pay rent, but many are simply squatters. A second category of Brazil's supply of informal housing includes ramshackle houses divided into several small separated units. These houses are locally known as "corticos," meaning "beehives." Many of the nine million inhabitants of informal housing in Sao Paulo, Brazil, occupy corticos. Interior walls and ceilings are constructed of weak materials such as cloth, cardboard, or plastic. In some areas only 54 percent of rooms have windows. Regardless of the type of slum-dwelling involved, overcrowding of occupants seems universal. Sanitation is universally inadequate, additional hazards include poor electrical and natural gas connections, where such facilities exist at all. Infestations of cockroaches, rats and similar biological hazards are common.

Situation is a major reason why some cities thrive and others decline. A city with a situation that nurtures its prosperity (a) will contain activities that compliment the enterprises in surrounding locations; (b) equivalent activities will not be available in other locations nearby; and (c) distance/transport time will not discourage spatial interaction between the city and its hinterland (Ullmann and Boyce, 1980). Situational advantages provide the basis for growth of major types of influential urban areas, which include gateway cities, primate cities and world cities.

Gateway Cities

Favorably situated cities become gateways; they provide access to their hinterlands for people, goods and capital. The prosperity of gateway cities depends on the prosperity of their hinterlands. If, for example, the hinterland consists of subsistence agriculture there is unlikely to be a demand to trade. Consequently, the nearby cities will lack, or not have, a gateway function. Yet, if the hinterland can provide a valuable resource such as oil, the city becomes the connection point for prospective oil traders. Edmonton, Canada, is an example of a gateway city for the oil fields of northern Alberta. In such instances, the gateway city can grow rapidly, but when the hinterland's valuable product or service is depleted the gateway city's prosperity declines. Ghost towns illustrate this phenomenon. The once-thriving city of Pithole, in Venango County, Pennsylvania, was erected quickly when commercial oil wells were opened in the immediate vicinity; Pithole's population rose to 20,000 in December 1865, but fell to 2,000 by December 1866 as the oil was depleted. Gateway cities of more enduring importance can be seen around the world. For example, Sydney, Australia and Vancouver, British Columbia are gateways for immigrants from the Pacific Rim to Australia and Canada, respectively; while air travel hubs exist in places such as Atlanta, Minneapolis, Paris and London.

Primate Cities

Perhaps the most centralized of all urban places are "primate cities"—i.e., disproportionately large and important cities of the country where they are located. Primate cities are at least twice as large as their country's second-largest city and they are also dominant in terms of social, economic and/or political influence. As a result of this imbalanced importance, goods, services, capital and migration flow from hinterlands into the primate city. Primacy can be economically efficient because national capital can be focused on meeting the needs of only one city. This may result in creation of one large system of police, fire, public transportation, etc., rather than duplicate systems for numerous cities of similar sizes. However, huge populations concentrated in one city can result in pollution, congestion and social problems on massive scales. Moreover, massive investment of national wealth into addressing conditions in primate cities can usurp resources needed for combating deterioration in other cities in a given country. The phenomenon of urban primacy is more pronounced and more likely to occur in LDCs than in MDCs. Examples of primate cities include Manila (Philippines), Jakarta (Indonesia) and Bangkok (Thailand), whose populations' social, political and financial importance are far more than double the next-largest city in each respective country. In countries with limited economic diversity, urban functions on a large scale are not regularly required—nor can they be regularly afforded—by most people. In addition, the largest city in a given LDC may provide the country's only non-agrarian economic opportunities. Concentration of political power and labor specialization can also encourage development of a primate city. When the population of a primate city grows, gross domestic product usually increases. However, this can create localized concentrations of wealth in LDCs. Of course, the growth of commerce comes with costs borne by individuals. Henderson (2002) compared cities located in various LDCs around the world to identify some effects of urban primacy. He showed that costs of housing and transportation were much higher in primate cities than in other cities in the same countries. For example, he calculated that rents and commuting costs could be expected to be 80 percent higher in a city of 2.5 million people compared with a city of 250,000 people.

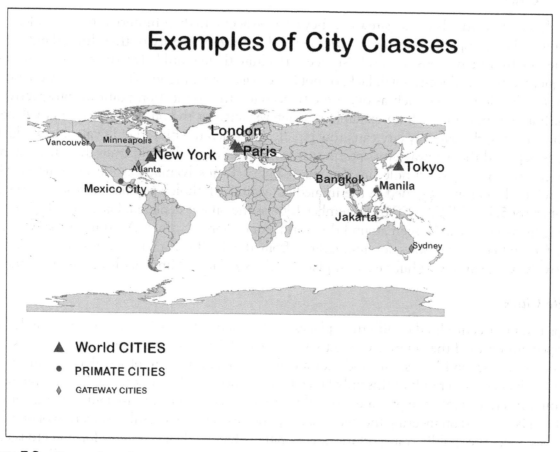

Figure 5.3   Examples of major urban types—gateway, primate, and world cities. (© 2011, Ronald Runeric)

World Cities

World cities support financial and business operations of world importance; this designation implies a major influence on the operation of global capitalism. Defining the concept of world cities engenders less debate than identifying examples. The importance of these cities' contributions to the fields involved is evaluated subjectively. Scholars generally agree that, at least, London, New York, Paris and Tokyo qualify as world cities according to acceptable standards. The major functions of world cities offer financial, legal and trading services that are interconnected with related operations in other parts of the world.

While global capitalism utilizes world cities as bases and distribution points from which to organize international operations, manufacturing leaves world cities for locations with minimal labor cost. For employees, middle-income opportunities are sparse, so most occupations are either high-wage or low-wage types. World cities are locations from which news and entertainment is broadcast to other regions, so they streamline the process of popularizing ideologies. Consequently the impact of world cities on the global social/economic order is staggering.

World cities are similar to gateway cities, because they are points from which business operators concentrate global business functions at a few points and channel inputs, outputs and especially, profits. The income generated by these operations is so coveted that many planners and officials of large urban areas aspire to nurture their jurisdiction's rise to the status of world city. A case in point is Tokyo, the world's most populous metropolis (13 million in the central city and 36 million in the metropolitan area). Japanese financial businesses are regulated by the federal Ministry of Finance so that they make money available to develop Japan's industries and economy. By emphasizing reinvestment and employment rather than high profits and individual consumption, Japanese policies have actively discouraged growth in the kinds of services distinguishing New York City (Hill and Kim, 2000). In contrast to other world cities, manufacturing remains a significant employer in Tokyo rather than being dispersed to other countries where labor cost is minimal.

URBAN ECOLOGY

During 2007, the world's population became more urban than rural. Metropolitan areas are sinks for huge quantities of the desirable products of nature, while generating large quantities of refuse and sewage. Large, concentrated populations use huge quantities of natural resources, including most of the world's fuels for manufacturing, electricity and transportation. Interactions with the natural environments near and within any metropolitan area are unavoidable and must be sustainable. Fresh water is normally obtained near each urban area and at least some food is generally produced nearby. Obtaining both necessities still requires plenty of input from nature. People in cities demand natural settings for their recreations. Yet, urban activities often disrupt operation of large or desirable ecosystems, sometimes with wide-ranging consequences. Human reactions to these disruptions range from indifference to consternation to anger.

The physical characteristics of cities are not always environmentally benign. Buildings and streets reduce groundwater recharge. Motorized vehicle traffic, electricity generation and cooling and heating buildings can affect climate and pollute air. People produce sewage and consumption of goods produces refuse. Urban areas break up natural habitats and these breaks become more numerous and larger as cities increase in size and merge with adjacent metropolitan areas. Large, concentrated and growing urban populations continuously intensify these problems and efforts to address them have often failed to keep pace.

We have already described cultural ecology in chapter one as the effect of human cultures on the natural environment. Since cities are collections of man-made environments, the study of urban ecology focuses on both natural and social concepts; it is an interdisciplinary study of these effects specifically in urban settings.

Population growth in any given city strains supplies of resources that the population needs or desires. Since supplies of many vital resources are finite, the number of people that a given city can support is clearly limited. The understanding of the concepts of carrying capacity and social carrying capacity are fundamental for studying urban ecology. While the ecological concept of carrying capacity is the maximum number of individuals that can be supported in a given area, the social carrying capacity emphasizes the number of people that can be sustained at a given economic standard of living. For example, cities without parks and other green spaces might be capable of supporting additional population, but this would entail a reduction of living standards that a community might not consider acceptable.

Urban Sprawl

Urban sprawl is defined as the unmanaged horizontal growth of urbanized space. It is widely regarded with concern for the following reasons:

- Sprawl utilizes valuable land that could be used for agriculture.
- Agriculture on remaining small patches of land near urban areas is difficult and less efficient than farming large acreages.
- Valuable natural environments are often disrupted or destroyed by urbanization.
- People value open space for comfort and healthful recreational activities.

Causes of urban sprawl are generally related to urban population growth, easy personal transportation, usually by automobile, desire of people to own at least some land in addition to a house and declining quality of urban life.

Sprawl is ongoing in most areas of the world, including both developed and developing regions. In the United States, the spatial extents of many small and medium metropolitan areas are expanding. For example, Centre County, Pennsylvania, long a largely rustic region featuring agriculture on fertile valley soils and forests on steep hills, now includes growing urbanization surrounding the Penn State University campus. In Gallatin County, Montana, exurban growth, i.e., suburban developments distant from associated urban cores, is ongoing. Gallatin County borders the west entrance to Yellowstone National Park and is included in the ecologically sensitive Greater Yellowstone Ecosystem. The county has experienced >30 percent increase in population during the 1970–1980 and 1990–2000 decades. Table 5.1 illustrates urban sprawl by comparing the areal extents of Cleveland, Ohio and Amsterdam, Netherlands. Amsterdam's central city population is 185 percent greater than Cleveland's central city population, in approximately the same area. The Cleveland metropolitan area is 286 percent larger than Amsterdam's metropolitan area, with approximately equal population. The central cities are approximately equal in size, yet Amsterdam's central city is home to 185 percent more people. The population of Amsterdam lives in a relatively compact area around the central city, whereas the population of Cleveland is dispersed over a larger area that is not completely urban, nor is it rural. It is merely a built-up environment.

Demands for resources across numerous and scattered, low-density suburbs may be environmentally and economically unsustainable. Agricultural land and level, wooded lands immediately beyond areas already suburbanized tend to be particularly attractive to developers of suburban hous-

Figure 5.4   Urban Sprawl. (© 2011, SeanPavonePhoto. Used under license from Shutterstock, Inc.)

	Central City Area	Central City Population	Metropolitan Area	Metropolitan Area Population
Cleveland, OH	213.4 km² (82.4 mi²)	478,000	7011 km² or (2707 mi²)	2,250,000
Amsterdam, Netherlands	219 km² (84.6 mi²)	1,364,422	1,815 km² (700.8 mi²)	2,158,372

TABLE 5.1   Comparing Cleveland, Ohio, and Amsterdam, Netherlands.

ing. Such areas are often unrestricted by legal restrictions on construction and usually only require elimination of natural vegetation before construction can begin. Burchell et al. (2002), estimated that 18.83 million acres of new development will occur in the United States between the years 2000–2025. Only 4.7 million of those acres will be land that is presently barren, unproductive and not environmentally fragile. Of the rest, 7.09 million agricultural acres and 7.04 million environmentally fragile acres will be consumed by these developments.

Urbanization and Climate

Urban areas are usually warmer than the surrounding countryside; these zones of excess heat are often called **urban heat islands.** Heat escapes from buildings, vehicles, sewers and people and their activities to enhance ambient temperatures in cities. The urban-to-rural temperature differential is greater during summer than winter and greater during summer days than nights. This difference is believed to be caused by the expulsion of warm exhaust of air conditioners into the atmosphere, whereas the heat generated in buildings during winter is intended to remain indoors. Rural-versus-urban temperature differences are also influenced by the natural climate and vegetation where cities are located.

Urban-related activities are major contributors to human-induced climate change, popularly called "global warming" and is based on the "Greenhouse Effect." The surface of the earth is heated by short-wave energy transmitted from the sun, which is reradiated as long wave energy back into the atmosphere. The atmosphere contains several greenhouse gases such as carbon dioxide (CO_2) that reflect this long wave radiation back to the surface, thus reheating the lower atmosphere and causing a rise in surface temperatures.

Many mundane human activities emit carbon dioxide into the atmosphere, increasing the amount of heat that can be retained in the lower atmosphere. The use of motor vehicles contributes heavily to greenhouse gas emissions. This is especially true in relatively wealthy countries, although car ownership and use in developing countries is also escalating. Travel in and between urban areas is a major contributor. As cities sprawl, inter-urban stop-and-go traffic intensifies. Additionally, new residential suburbs are often developed quite distant from established urban workplaces.

Urban sprawl is also associated with other air quality disturbances, such as low-level ozone (O_3). Ozone is formed when volatile organic compounds (VOCs) and nitrogen oxides (NOx) are exposed to heat and sunlight. Both VOCs and NOx are products of combustion processes that occur on large scales in cities, such as driving, lawn mowing and combustion related to industries. During hot, still summer weather, elevated concentrations of low-level ozone are common. While ozone in the stratosphere protects terrestrial animals from excessive exposure to ultra-violet energy from the sun, ozone is harmful at the earth's surface. Plant growth and survival can also be impaired due to ozone exposure and long-term cumulative impacts on trees are believed possible (USEPA, 1997). Even healthy children and adults can experience a reduction in respiratory efficiency during periods of elevated levels of low-altitude ozone.

Urbanization and Waste Disposal

One of the major problems associated with urban areas is the large accumulation of waste products, whether in the form of sewage, or solid waste—garbage or trash—it has to be treated in some form or another so that it is not a hazard to public health.

Raw or treated sewage is mostly water and will eventually be returned to surface waters. If treatment has been successful, the disruption of ecological processes in the receiving waters will be held to some acceptable level. During most of human history, untreated sewage customarily was directed into streams, lakes, or oceans without treatment and this practice is still common in some areas of the world. For example, most Brazilian cities have no sewage treatment. In Sao Paulo state, only 16 percent of sewage is treated at all. The rest is dumped directly into streams and preferentially into small streams.

Modern sewage treatment involves three major stages. Each of the three stages produces some solid material (called sludge if it consists of small particulates in a mass); disposal methods depend on the content of the solids. Burial in landfills, sale as fertilizer, or drying and burning are options, depending on the composition of the sludge. In the United States in 2004, over 21,000 publicly-owned sewage treatment facilities served 78 percent of the population. The United States Environmental Protection Agency (USEPA) has established standards for acceptable levels of contaminants in wastewater. Many U.S. municipalities have combined sanitary and storm sewers; these are systems of pipes that gather household and commercial sewage plus storm runoff into a single system of pipe that delivers all wastewater to the town's treatment facility. Although these systems are designed to accommodate large quantities of wastewater, occasional major storms can overwhelm the system and wastewater must bypass the treatment facility and flow untreated into local surface waters. During floods, the sewage may also flow untreated into homes. This problem is not easily solved; although sewers are generally designed to provide an efficient system even during unusually heavy precipitation and/or runoff, this must be balanced against the cost of designing a foolproof system.

Solid waste, garbage, trash and solids removed from wastewater must be recycled, burned or buried. However, in much of the world, open dumps still receive most refuse. For example, the Bangkok, Thailand, metropolitan area produces 2.4 million tons of solid waste annually, most of which is ultimately placed into open dumps. The growth in the amount of solid waste has not been in proportion to the growth in population. In the United States, annual production of solid waste is about 250 million U.S. tons (227 million metric tonnes) annually; this is approximately 0.82 tons (0.74 metric tonnes) per capita (USEPA, 2008). Per capita creation of refuse is six times greater in the United States than in the rest of the world.

In recent years across the developed world, there has been a realization that much of the solid waste produced can be recycled. It is estimated that 60 percent of all household waste could be recycled but this is not always happening. In the United States, recycling has risen from 10 percent of refuse in 1980 to 28 percent today. Recycling of specific materials has grown even more drastically: 42 percent of all paper, 40 percent of all plastic soft drink bottles, 55 percent of all aluminum beer and soft drink cans, 57 percent of all steel packaging and 52 percent of all major appliances are now recycled (USEPA, 2008).

About 32 percent of municipal solid waste in the United States is incinerated and the heat is used to generate electricity (USEPA, 2008). Incineration can be environmentally desirable for dangerous refuse, such as hospital waste. Dangerous pathogens are destroyed by burning; in addition, incineration compacts the refuse that must ultimately be interred in landfills. However, toxic ashes and hazardous atmospheric effluents can be generated, depending on the chemical composition of the refuse. The hazardous ash must be buried in a landfill designed to secure toxic wastes.

Landfills are the ultimate destination for over 50 percent of solid waste in the United States, although this figure is down from approximately 81 percent in 1980. Municipal and industrial solid waste that cannot be recycled or burned is buried in landfills. Until the second half of the 20th century, most municipal solid waste was transported to open city dumps, where refuse was often burned. During the last 30 years, however, governmental regulations have required establishment of improved disposal facilities. Modern sanitary landfills are designed to safely retain refuse and by-products of decomposition on-site. However, problems associated with concentrations of widely-disparate types of refuse are inevitable. Capacity remains adequate, though, because the remaining landfills tend to be considerably larger than those of previous decades.

Urbanization and Water Quality

Like many other human activities, urbanization often affects the quality of surface waters by impairing or contaminating either runoff or groundwater. Cities impact the supply and quality of nearby surface waters, i.e., streams, lakes and oceans. Roads and buildings are impermeable, so the natural process of water percolation into the ground after rain or snowmelt events is impossible. Impermeable surfaces generally increase the rate of runoff and decrease the rate of evaporation. Thus, surface waters in urbanized areas can rise quickly after precipitation or snowmelt, even to the point of flash flooding.

In urbanized areas there is an increase in the use of many chemical products that eventually enter streams and lakes via runoff, or percolate into groundwater which feeds the surface waters. Lawn and garden products, such as fertilizers and weed killers, can wash into natural waters after heavy rains or during snowmelt events. Constituents of these and other common consumer products increase the quantities of chemicals often found naturally in many surface waters. In excessive concentrations, chemicals can become problematic by promoting excessive growth of aquatic plants, or, at extremely high concentrations, can be toxic to some aquatic life.

Stream ecosystems are highly vulnerable to major ecological disruption by both urbanization and highway construction. Major construction programs result in flows of runoff carrying great quantities of fine sediment into nearby watersheds. Large quantities of fine sediment can clog gills of small fish. When it settles to stream beds, it can smother eggs of aquatic biota and thus reduce the forage available for fish and other animals. Once highways are open for use, driving becomes a further source of water quality impairment. Chemical and heavy metal residues from operation of motor vehicles are subsequently transported by runoff into streams and lakes. So much driving-related pollutants occur along roadsides that it is estimated that in some U.S. locations, quantities may be sufficient to support profitable mining of roadside soils.

The Impact of Urbanization on Habitats

Since the end of World War II, urbanization has impacted natural forest areas in many parts of the world. Residential development in previously-forested areas not only reduces the total quantity of forest habitat for flora and fauna, but the development of housing and roads within wooded areas creates fragmented habitats. Natural habitats become so interspersed with construction that some plants and wild animals may be unable to survive. Some flora or fauna may require relatively large amounts of contiguous habitat, so much so that the patches of natural habitats remaining after residential development may not be sufficient for their survival. Identification of species that can thrive in fragmented habitats and those that cannot has become a major area of ecological study.

It must be stressed that habitat fragmentation does not always lead to extinction, even for predatory species that utilize relatively large territories. Most studies of habitat fragmentation conclude that fragmentation limits animal populations. For example, long-distance migrant bird species tend to avoid areas with high levels of road density and urban land cover. Fragmentation also imposes survival difficulties on "habitat specialists," i.e., animals that are not able to adapt to various habitats. An example from Western Australia identifies two species of woodland gecko impacted by development. The reticulated velvet gecko is a habitat specialist and the tree dtella (house gecko), is a habitat generalist. Over 90 percent of the specialist gecko's original habitat has been eliminated by forest clearing. Although the species has survived in some of the forest patches that remain, these geckos are absent from other patches where they almost certainly lived before their habitat was fragmented. On the other hand, the tree dtella, the habitat generalist gecko, appears to thrive in the fragmented habitat.

Implications for the Future

Urban-related environmental problems persist probably because they are large, difficult and expensive to solve, rather than because people are indifferent to the natural environment. Problems related to urban growth can sometimes be addressed individually, but these possibilities are limited. As long as human population grows, urban areas probably will grow also. For the foreseeable future, continued pollution and destruction of forests and natural habitats are likely; concurrently, demand for high-quality natural environments will probably also increase. In the future, the amount and quality of unimpaired or restored natural environments proximate to urban settings will reflect societies' commitment to meeting those demands.

KEY TERMS

Site

Situation

Urban Heat Island

Gateway cities

Primate cities

World cities

Formal/informal housing

Favela

Cortico

Habitat fragmentation

QUESTIONS

1. Why is the "urban export function" important for the economic strength of a city?
2. What are break-of-bulk points, and how do they influence urban growth?
3. Explain the major differences between sanitary landfills and open dumps.
4. Describe "highway runoff" and its potential consequences.
5. Explain urban sprawl and relate this phenomenon to environmental concerns such as those you mentioned when answering questions 3 and 4 above.
6. How has urbanization stressed natural environments in MDCs compared with LDCs?

CHAPTER 6

The Diversity of Language

It can be said that language is the single most important cultural trait. Language is a set of words or actions, together with the methods of combining them, that is used and understood to communicate within a group of people. In this definition we also include sign language and body language. Each language is a unique way of dealing with facts, ideas and concepts. Variations in language lead to differences in how people think about time and space and about how thoughts regarding various actions are processed. Exact translation from one language to another is virtually impossible. The language an individual speaks influences the structuring of his or her perception and ways of thinking.

The term language is usually reserved for major patterns of differences in communication. However, many groups of people who communicate exclusively with one another will soon develop a jargon—"their own language"—whether they are nuclear physicists, a social group or high school students. Communications can be fraught with difficulties when translating between two different languages. For example, it would be ridiculous to sell a car in Mexico called a Nova whose plural is Novas (which in Spanish is no go). Problems can also arise with English speakers from different parts of the English speaking world where the same word can have very different meanings. A person from Britain asking for chips in an American restaurant would be surprised to receive potato chips, when they really should have asked for French fries. Similarly, requesting a biscuit would not, for a Brit, result in being proffered a cookie but a dinner roll. British ladies may have ladders in their stockings while her American counterpart would have a run in her hose or tights. Using words or phrases that have totally different meanings in the two countries can also lead to embarrassing situations; for example, the British colloquialism for a cigarette is a fag, which in the United States has an entirely different meaning.

Another problem that can occur is that many languages have regional variations in the word use. Minor variations within a language are called **dialects**—regionally distinctive ways of pronunciation, usually with a set vocabulary. For example, a term of endearment in the English midlands is ducks as in "how are you ducks?" where a person in Lancashire would use chuck in place of ducks, and a Glaswegian may use hen. In the United States, the modern use of Ebonics (African American vernacular English) can also be regarded as a dialect.

THE DISTRIBUTION OF LANGUAGES

The great variety of languages spoken today testifies as to the relative isolation of different culture groups in the past. The distribution of any language illustrates the pattern of dispersion of its original speakers or their cultural impact on others. There are over 6,000 different languages spoken in the world today. Three hundred eighty-nine (or nearly 6 percent) of the world's languages have at least one million speakers and account for 94 percent of the world's population, while the remaining 94 percent of languages are spoken by only 6 percent of the world's people (Lewis, 2009). Over half the world's peoples speak one of 12 major languages. See Table 6.1.

In contrast to these widespread languages other languages are extremely local. Linguists have discovered fully developed languages in New Guinea spoken by only a few hundred people living in isolated valleys. These languages have developed in total isolation over long periods of time and are totally incomprehensible to other peoples living only a few miles away in the next valley.

The actual number of different languages varies with the accepted definition of language. The complications arise when cultures meet and have a need to communicate. When different linguistic groups come in contact with each in order to aid communication, a new common language is derived. This is called a **pidgin language** and is characterized by a very small vocabulary, which is a combination of the languages of the groups in contact. An example of pidgin is found in New Guinea, for instance *gut bai* (good bye) and *tenkyu* (thank you) reflect the influence of English. When pidgin languages acquire their own vocabularies and become native languages of their speakers they are called **Creole** languages. These are linguistic variations derived from Indo-European languages initially; Spanish and French Creole developed in the West Indies, English-based Creole developed in many of the former British colonies. There is also an Arabic Creole that has developed in northern parts of sub-Saharan east Africa, notably Kenya and Uganda.

With increasing amounts of trade and communication there becomes a need for one language to dominate. This is known as a **lingua franca,** or a language of communication and commerce over an area where it is not the mother tongue. The Swahili language enjoys lingua franca status in much of East Africa, while English has become the global lingua franca.

An interesting example of the development of a lingua franca is Hawaiian Pidgin, which began as a pidgin jargon used in the early European colonization of Hawaii. Although English served as the main language, there were many Chinese, Japanese and Spanish people intermixed with the native Hawaiians. These people started using Hawaiian Pidgin as a lingua franca, and by the mid-1900s it had creolized to become a minor language of Hawaii, and remains so today.

One final note: a country where there is a mixed population with many different languages is known as a **polyglot,** or multilanguage state, e.g., the United States. Some regions are characterized as **bilingual,** which is the ability to speak two or more languages fluently. Because second languages are usually learned later than first languages, bilingualism is usually not uniform across a community (Lewis, 2009). Where cultural groups use a second language, there are usually varying degrees of bilingual proficiency, ranging from the ability to only use as a greeting, to being able to freely communicate in the second language. Cultural leaders, the educated, traders, those who travel extensively and people in large urban areas, are liable to be more bilingual than others in the group.

Language	Family	1st Language Speakers in Millions	Total Speakers	Main Areas Where Spoken
Mandarin	Sino-Tibetan	1,080	1,460	China, Taiwan, Singapore
Hindi	Indo-European	370	496	Northern India
Spanish	Indo-European	358	425	Spain, Latin America
English	Indo-European	322	514/1,000*	British Isles, North America, the Philippines, former British Colonies
Malay-Indonesian	Austroesian	223	250	South East Asia
Portuguese	Indo-European	210	230	Portugal, Brazil, African colonies
Arabic	Semitic	206	254	Middle East, North Africa
Bengali	Indo-European	171	215	Bangladesh, Eastern India
Russian	Indo-European	145	255	Russia, Kazakhstan, Ukraine, former Soviet republics
Japanese	Japanese-Korean	127	128	Japan
French	Indo-European	109	239	France, French overseas territories, Switzerland, Belgium
German	Indo-European	100	122	Germany, Austria, Switzerland, Luxembourg, Northern Italy

This number varies with basis of category. Some scholars recognize it as being close to 1 billion.

TABLE 6.1   Major Languages Spoken as First and Second Languages. (U.S. Department of State)

LANGUAGE FAMILIES

The mosaic of languages across the globe reflects the long, turbulent history of humankind. The distribution is complex, encompassing literally thousands of dialects and languages, each spoken in its own distinct area. Some order can be brought to this seeming chaos if we recognize that most languages belong to families consisting of related tongues derived from a common ancestral speech. In this chapter we discuss nine major language families plus Uralic that account for 95 percent of the world's population.

Language Family	Count[1]	Number of Speakers		Locations
			%	
Indo-European	426	2,721,969,619	45.67	Europe, North America, South Asia, South America, Belarus, China, Fiji, Iran, Iraq, Israel, Lithuania, Maldives, Oman, Russian Federation, South Africa, Suriname, Tajikistan, Turkey, Ukraine
Sino-Tibetan	445	1,259,227,250	21.13	Bangladesh, Bhutan, China, India, Kyrgyzstan, Laos, Myanmar, Nepal, Pakistan, Thailand, Vietnam
Niger-Congo	1,510	382,257,169	6.41	Sub-Saharan Africa and Southern Sudan
Afro-Asiatic	353	359,495,289	6.03	North Africa, the Middle East, Cameroon, Cyprus, Georgia, Iran, Kenya, Malta, Nigeria, Somalia, Tajikistan, Tanzania, Turkey, Uzbekistan
Austronesian	1,231	353,585,905	5.93	Indonesian Archipelago, The Pacific Islands, Southeast Asia, Chile, China, Madagascar, Myanmar, New Zealand, Papua New Guinea, Philippines, Suriname, Taiwan
Dravidian	84	222,682,100	3.74	India, Nepal, Pakistan, Sri Lanka
Altaic	64	139,525,936	2.34	Afghanistan, Azerbaijan, China, Georgia, Iran, Kazakhstan, Kyrgyzstan, Moldova, Mongolia, Russian Federation, Turkey, Turkmenistan, Ukraine, Uzbekistan
Japonic	12	123,090,950	2.07	Japan
Austro-Asiatic	169	103,703,873	1.74	Bangladesh, Cambodia, China, India, Laos, Malaysia, Myanmar, Thailand, Vietnam
Totals	4,329	5,686,953,131	95	

1 = The number of living languages in a language family.

TABLE 6.2   Major Language Families. (Lewis, 2009)

In 1786 Sir William Jones, a British judge stationed in India, recognized the relationship between Sanskrit, Latin and Greek, as we shall see later. His work led to the scholarly acceptance of language families, and during the first half of the 19th century, many languages were investigated by Western scholars who created word lists and grammatical sketches. The comparative method for determining genetic relationship among languages was worked out in detail for the Indo-European language family during the 19th century. The methodology takes sound associations in related words and morphological units, along with structural similarities in phonology, morphology, and syntax. Some of these are explained in terms of a reconstructed common language, or protolanguage. Structural similarities in many cases are due to interaction among contiguous languages over a long time creating linguistic areas.

The Indo-European Language Family

The largest and most widespread language family is the **Indo-European,** which is dominant in Europe, Russia, North and South America, Australia and parts of southwestern Asia and India. Subgroups such as Romance, Slavic, Germanic, Indic, Celtic and Iranie are part of the Indo-European family, and they in turn are subdivided into individual languages. For example, English is a Germanic Indo-European language.

If we compare the vocabularies of various Indo-European tongues, we can readily see the kinship of these languages. For example, the English word *mother* is similar to the Polish *matka*, the Greek *meter*, the Spanish *madre*, the Armenian *mair*, the Avestan (spoken in Iran) *matar*, and the Sinhalese (spoken in Sri Lanka) *mava*. Such similarities in vocabulary suggest that these languages had a common ancestral tongue. We will discuss this language family in more detail later in the chapter.

Sino-Tibetan Family

After the Indo-European language family, the Sino-Tibetan group comprises the world's second largest language family and includes 445 languages and dialects. There are three major divisions within Sino-Tibetan; Sinitic, Tibetic and Burmic, plus the Karen Language, while some scholars include Tai and Hmong-Mein. Chinese, the name given to the language that has been applied to numerous dialects, styles, since the middle of the 2nd millennium, is a misnomer. Sinitic is the correct designation covering all these languages although *Han* is a term used in China to distinguish Chinese from non-Chinese languages. The Chinese terms for Modern Standard Chinese are *Putonghua* "common language" and *guoyu* "national language" (the latter term is used in Taiwan).

Sinitic languages are spoken in China, Taiwan and in all the countries of Southeast Asia. They have also spread by immigration to Oceania and North and South America. In total there are at least 1.14 billion Chinese speakers in the world today. Sinitic is divided into a number of language groups, the most important of which is Mandarin (Northern Chinese) based on the dialect of Beijing. This is known as Modern Standard Chinese, having been adopted as the official form of speech for the People's Republic of China. Mandarin is not only the most important language of the Sino-Tibetan family but also has the most ancient writing tradition of any modern language. Other Sinitic language groups include Wu, found in the eastern provinces around Shanghai and spoken by about 85 million people; Yue, or Cantonese found in Guangdong and Hong Kong; and Xiang in Hunan province and Taiwanese.

Tibeto-Burman languages are spoken in the Tibet Autonomous region of China and Myanmar (Burma); in the Himalayas states of Nepal and Bhutan plus areas of northern India, Pakistan, and Bangladesh; they also are spoken by hill tribes throughout Southeast Asia and central China. The Tibetan and Burmese writing systems are derived from the Indo-Aryan (Indic) tradition but have in recent times acquired writing systems in Roman script or in the script of their host country—Thai, Burmese and Indic, for examples.

The structure of the Sino-Tibetan languages is for the most part rather simple and nonspecific. A number of features are common throughout. Many of them can be shown to be of a typological nature, the result of diffusion and underlying unrelated language strata. The vast majority of words in all Sino-Tibetan languages are, with few exceptions, one syllable. Also, most languages in the group possess phonemic tones, which indicate a difference in meaning in otherwise the same word; for example, *ma* can be pronounced in several different ways and have as many meanings.

Niger-Congo Family

The Niger-Congo is a family of languages of Africa, which is by far the largest language family in Africa. The area in which these languages are spoken stretches from Senegal, at the westernmost tip of the continent, east to Kenya, and south to South Africa. Over 600 million people or 85 percent of the population of the continent speak a Niger-Congo language. In some countries, such as Niger and Chad, these are minority languages, while in most African countries they are in the majority. It is estimated that the number of Niger-Congo languages is about 1,400. The greater part of the Niger-Congo culture region belongs to the Bantu subgroup, which includes Swahili, the lingua franca of East Africa. Both Niger-Congo and its Bantu constituent are highly fragmented into a great many different languages and dialects. Some languages have a special dialect such as Swahili, which has 17 separate dialects and 15 additional variant names for some of the dialects.

The name Niger-Congo was coined to reflect the predominance of these languages in the great river basins of the Niger and Congo rivers. The languages of the present-day Niger-Congo family are divided into nine major branches: Mande, Kordofanian, Atlantic, Ijoid, Kru, Gur, Adamawa-Unangi, Kwa, and Benue-Congo. These nine branches relate to each other in different ways, some being closer to each other than others. Adamawa-Ubangi and Gur, for example, appear to be closer to each other than, say, Kru and Kwa. These varied relationships suggest the nine major branches did not derive directly from a common ancestor but developed by diffusion over time.

A system of noun classes is probably the characteristic most widely found in Niger-Congo languages. The extent to which the system operates varies greatly, but is found in some form in each of the languages of the family. Most systems use some form of prefix or suffix to distinguish their singular and plural forms. Most Niger-Congo languages have a tonal system, commonly with two or three contrasting levels of pitch (though four levels are also found and very occasionally even five). The function of tone varies from language to language; sometimes it marks a grammatical feature, other times a word contrast. There are also about 10 vowel sounds, although not all languages use them.

The Afro-Asiatic Family

The **Afro-Asiatic Language family** consists of two major subdivisions, **Hamitic and Semitic.** These two groups originated in southwestern Asia before recorded time.

Hamitic languages are spoken almost exclusively in Africa, by the Berbers of Morocco and Algeria, the Tuaregs of the Sahara, and the Cushites of East Africa. The Hamitic speech area was formerly much larger than it is now. It once covered the lands of the ancient Egyptians, but it was greatly reduced and fragmented by the expansion of Arabic over a thousand years ago.

The Semitic languages cover the area from the Arabian Peninsula and the Tigris-Euphrates river valley in the Fertile Crescent of Iraq westward through Syria and North Africa to the Atlantic Ocean. Despite the considerable size of this domain, there are fewer speakers of the Semitic languages than of the other major language families, mainly because most of the areas Semites inhabit are sparsely populated deserts. The ancient Babylonians, Assyrians, Phoenicians, and Hebrews were Semites

We can identify two major subgroups of the language family, North Semitic and South Semitic; two languages, Aramaic and Hebrew, belong to the former group. Aramaic, the language of a people living in the western edge of the Syrian Desert, became predominant over a wide area in the Middle East, most probably because of the extensive commercial relations developed by the Aramean states. Aramaic dialect was the vernacular of Palestine during the time of Christ and at least one of the Gospels can be considered as pure Aramaic literature.

Aramaic also had a considerable effect on certain languages spoken in Iran and traces are still apparent in the northwest of the country. The language itself lingers in a few villages near Damascus and Mosul and is the sacred language of a number of Christian sects in Syria and Iraq.

Hebrew also is a Semitic tongue, closely related to Arabic. For many centuries, Hebrew was a "dead" language, used only in religious ceremonies by millions of Jews scattered around the world. With the creation of the state of Israel in 1947, a common language was needed to unite the immigrant Jews, who spoke the languages of many different countries. Hebrew was revived and made the official national language. Hebrew had lain dormant for 2000 years. It had to be modernized. To make the transition to the twentieth century, words had to be coined for *telephone, airplane, rifle,* etc.

العربية

Arabic is by far the most widespread Semitic language and has the greatest number of speakers, about 235 million. Although many different dialects of Arabic are spoken, the written form is standard. Arabic ranks sixth in the world's league table of languages, with an estimated 186 million native speakers. As the language of the Qur'an, the holy book of Islam, it is also widely used throughout the Muslim world.

There are many Arabic dialects. **Classical Arabic**—the language of the Qur'an – was originally the dialect of Mecca in what is now Saudi Arabia. **Modern Standard Arabic** is used in books, newspapers, on television and radio, in the mosques, and in conversation between educated Arabs from different countries (for example at international conferences). **Local dialects** vary, and a Moroccan might have difficulty understanding an Iraqi, even though they speak the same language.

Migrants from southwestern Arabia brought Semitic speech to Ethiopia about 3,000 years ago. There, in the isolation of the East African mountain highlands, it gradually evolved into Amharic, a third major Semitic tongue, today claiming 18 million speakers.

Austronesian Family

Formerly called Malayo-Polynesian languages, the Austronesian family contains over 1,200 languages spoken by more than 350 million people in Indonesia, the Philippines, Madagascar, the central and southern Pacific island groups, parts of mainland Southeast Asia, and the island of Taiwan. In terms of the number of its languages and of their geographic spread, the Austronesian language family is one of the world's largest, only the Niger-Congo family of Africa has a greater number of languages.

Prior to European colonial expansions, Austronesian languages were the most widely distributed of any language family, extending from Madagascar off the southeast coast of Africa to Easter Island west of Chile, taking in 206 degrees of longitude. A majority of the languages are found within 10 degrees of the Equator, although some extend well beyond this, reaching as far north as 25° N in northern Taiwan and as far south as 47° S on New Zealand's South Island.

Despite the enormous geographic extension of the Austronesian languages, the relationship of many of the languages can easily be determined by an inspection of such basic subsystems as personal pronouns or the numerals. Apart from the 14 languages spoken by the aboriginal population belonging to the Formosan group found on the island of Taiwan, the family can be divided in to two distinct groups: Western and Central-Eastern Malayo-Polynesian.

The Western Malayo-Polynesian group includes Javanese, by far the largest single language within the family, with 85 million speakers. Malay with around 40 million speakers is one of the 118 languages of the group found in Malaysia; most of which are found on the islands of Borneo and Sumatra, while on mainland Southeast Asia and the island of Hainan there are only eight. Also included in this group are the 160 native languages of the Philippines. The island of Madagascar, far to the west of all the other major languages, is an interesting outlier. Settled by immigrants from southeastern Borneo over 1,000 years ago, it has a single language—Malagasy—with about 20 dialects.

The 900 languages of the Eastern Malayo-Polynesian group are equally divided among the remaining parts of Indonesia such as the island of New Guinea and Oceania, which comprises nearly all the languages of Polynesia, Micronesia and Melanesia.

Despite the enormous geographic extension of the languages, the relationship of many of the languages can easily be determined by an inspection of such basic subsystems as personal pronouns or the numerals. Generalizations are difficult because of their enormous number and diversity, words tend to be disyllabic (having two syllables), while vowels and consonants tend to be limited, especially in Polynesian.

Of all the Austronesian languages, Malay has had the most cultural impact. After the introduction of Islam at the end of the 13th century, sultanates were established in the Malay-speaking region of the Malay Peninsula and on the coast of northwestern Borneo in Brunei. In other areas, such as northern Sumatra, the southern Philippines, and the northern Moluccas, Islamic sultanates made use of local languages; however, the large number of Malay words in these languages suggests that Malay-speaking missionaries must have influenced the development of the cultures. Malay also became an important language of trade, a lingua franca when much of the China–Europe trade passed through the Strait of Malacca. It also developed the same status in the Indonesian archipelago, a role that it still has to this day.

The Dravidian Family

Dravidian (drəvĭd`ēən) is a family of about 23 languages that appears to be unrelated to any other known language family. The Dravidian languages are spoken by more than 200 million people, living chiefly in south and central India and Northern Sri Lanka. There are four sub-groups as in the map in Figure 6.1. The majority of Dravidian speakers speak one of four languages: Kannada, with over 40 million speakers; Malayalam, with 35 million; Tamil, with almost 70 million speakers; and Telugu, with over 70 million speakers. A minor group, Brahui, with close to a million speakers, is found in Baluchistan, Pakistan.

It is thought that the Dravidian tongues are derived from a language spoken in India prior to the invasion of the Aryans c. 1500 B.C. There are many words of Indic origin in the Dravidian languages, which in turn have contributed a number of words to the Indic tongues. The Dravidian languages have their own alphabets, which go back to a common source that is related to the Devanagari alphabet used for Sanskrit. While Brahui uses an Arabic script, Dravidian languages are noted for retroflex and liquid sound types. A distinctive feature is the formation of a comparatively large number of sounds in the front of the mouth. Verbs have a negative as well as an affirmative

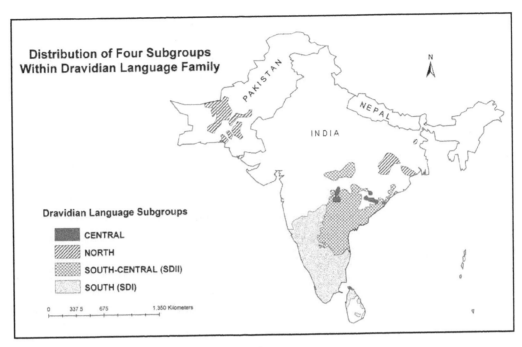

Figure 6.1   Dravidian Language Distribution. (© 2011, Leonard Peacefull)

voice. Gender classification is made on the basis of rank instead of sex, with one class including beings of a higher status and the other beings of an inferior status (to which inanimate objects and sometimes women are assigned). Nouns are declined, showing case and number. Great use is made of suffixes (but not of prefixes) with nouns and verbs.

Altaic Family

The Altaic language family consists of three branches: Turkic, Mongolic and Manchu-Tungus that show similarities in vocabulary, morphological and syntactic structure and certain other features, which are generally considered to be genetically related. The Altaic homeland lies largely in the inhospitable deserts, tundra, and coniferous forests of northern and central Asia. This group contains more than 50 languages, spoken by more over 135 million people spread across the breadth of Asia, from the Arctic ocean in the north to the latitude of Beijing about 40°N. The Turkic languages are found in a continuous band from Turkey, Armenia and Azerbaijan through the Central Asian republics of Kazakhstan, Uzbekistan, Turkmenistan, Kyrgyzstan and Tajikistan to Xinjiang in China. The Mongolian languages are concentrated in the adjacent, roughly oval region formed by Buryatia, Mongolia, and the Inner Mongolian province of China. The Manchu-Tungus languages are widely dispersed across eastern Siberia in Russia and Northeast China.

The Altaic peoples were historically concentrated on the steppes of Central Asia in or near the region of the Altai Mountains that form the western boundary of Mongolia with its neighbors Russia and Kazakhstan. The Turks inhabit the western, the Mongols the central, and the Manchu-Tungus peoples the eastern portions of the Altaic region. The diffusion of the language occurred as a series of migrations to the west and south. These migrations were partly a consequence of the economics of nomadic culture and partly due to conquest of neighboring sedentary populations. The states they founded, however, tended to be impermanent, often resulted in their eventual expulsion or in cultural and linguistic assimilation, as happened to the Manchu in China. The Turks, on the other hand, not only created a series of empires but formed the mass of the armies of the numerically inferior Mongol people, whose medieval empire was, outside of China and Mongolia, heavily Turkicized. These various developments left their mark in the vocabularies of the Altaic languages, though to a far lesser extent in their grammatical structures.

Japonic/Japanese Family

Japanese is one of the world's major languages. With 123 million speakers it ranks ninth in world languages. It is primarily spoken throughout the Japanese archipelago with another 1.5 million Japanese immigrants, mainly in North and South America. Of all the major language families, Japanese is the only one whose ancestry is in doubt. Speculation about it being part of other language families has continued over a long period. Korean remains the strongest ancestral language but others include Austronesian, the Austro-Asiatic, and the Tibeto-Burman languages.

The Yayoi Culture established in Kyushu about 2000 years BP is thought to have been introduced to Japan from the Asiatic continent. This culture group is thought to have brought a language of southern Korea, introduced iron and bronze implements and the cultivation of rice. This migration from Korea did not take place on a large scale, the new language did not eradicate certain older words, though it was able to change the structure of the existing language. Although it contains words linked to Austronesian it is theorized that Japanese is genetically related to Korean and, ultimately, to the Altaic language family.

The country's geography, consisting of high mountains, deep valleys, and small isolated islands, has fostered the development of various dialects, that in some case may be regarded as language subgroups. The Ryukyuan group of the southern chain contain 11 languages. The main islands have three distinct dialectic groups: Eastern, Western, and Kyushu. These differences of dialects create communication problems when two dialects are often mutually unintelligible. For example, the Kagoshima dialect of Kyushu is not understood by the majority of the people of the main island of Honshu. Likewise, the northern dialect of such places as Aomori and Akita are not understood in metropolitan Tokyo or western Japan. Today there is a standardized language or *kyōtsū-go*, a feature of compulsory education mobility and mass media. Culturally these processes helped to demolish dialectal differences with the resultant accelerated loss of local dialects.

Austro-Asiatic Family

The Austro-Asiatic family is a group of 169 languages spoken by more than 103 million people scattered throughout Southeast Asia and eastern India. Most of these languages have numerous dialects. There, three culturally significant languages within the family are Khmer, Mon and Vietnamese, all of which have the longest recorded history. The remainder are essentially a nonurban minority language that until recently were not written.

Superficially, there seems to be little in common between the languages. Vietnamese, for example, is a monosyllabic tonal language whereas Mundari? of India is a polysyllabic toneless language. Scholars using linguistic comparisons, however, confirm the underlying unity of the family. Relationships with other language families have been proposed, but, because of the scarcity of reliable data, it is very difficult to validate any theories. The work of classifying the members of the Austroasiatic family is in the initial stages. Historically, classification was based on geographic location. For example, Khmer, Pear and Stieng, are all spoken in Cambodia and, thus were all lumped together when they actually belong to three different branches of the Mon-Khmer subfamily.

Culturally, Khmer and Vietnamese are the most important languages in terms of numbers of speakers. They are the national languages of Cambodia and Vietnam, respectively. In each country the languages are taught in schools and used in mass media and in official business. Nearly all of the Austroasiatic languages have only recently been written, less than a century in most cases. Dictionaries and grammars have been written only for the most prominent languages while many languages have only been described briefly in a few articles, others are little more than names on the map. Outside of these two main languages, speakers of other Austroasiatic languages are under strong social and political pressure to become bilingual in the official languages of the nation in which they live. Most of these groups are too small or too scattered to win recognition, and for many the only chance of cultural survival lies in retreating to a mountain or jungle vastness, a strategy that reflects a long-standing Austroasiatic tradition.

Other Language Families and Groups

There are over 100 other minor language families that make up the remaining 5 percent of the world's spoken languages. They range from substantial groups such Tai with 80 million speakers to Zaporan with just 91 speakers. Even smaller numbers are found when one looks at actual languages, not families where there may be only one or two speakers still surviving. These groups, often occupy refuge areas after retreat before rivals, are remnant language families such as Khoisan, found in the Kalahari Desert of southwestern Africa and characterized by distinctive clicking sounds. Other smaller language groups include: Tai, Australian Aborigine, Papuan, Nilo-Saharan, Inuit and a variety of American Indian languages. In Europe the Basque language of Northwest Spain and the adjacent Southern corner of France is a unique language and related to no other family.

Uralic Family

One of the minor language families is one that has occupied a small part of the earth's surface in the tundra and grassland areas adjacent to the Altaic peoples. The Uralic family consists of more than 20 related languages, which have descended from a language that existed 7,000 to 10,000 years ago in the Ural mountains of Russia. Today these languages are spoken by more than 25 million people scattered throughout northeastern Europe, northern Asia, and North America. Demographically, the most important Uralic language is Magyar, the official language of Hungary. The language family consists of two related groups, the Finno-Ugric and the Samoyedic, which have within them divergent subgroups.

The Finno-Ugric languages are scattered over an immense Eurasian territory. In the west they include Hungarian, Finnish, and Estonian and Sami (or Lapp). Then there are those of the former Soviet Union extending from the Gulf of Riga to the Kola Peninsula. These include Estonian, Livonian, Votic, Karelian and Veps. The Mordvin and Mari are found in the central Volga region of Russia, while extending northward along river courses west of the Urals are the Permic languages; Udmurt, Komi (Zyryan) and Permyak (or Komi-Permyak). The easternmost outpost of these languages, Mansi and Khanty, are found in the Ob river basin.

Although there is a recognizable correspondence between these language groups there are great variances—so much so that the degree of similarity between some, for example, Hungarian and Finnish, is comparable to that between English and Russian (which belong to the Indo-European family of languages). The difference between any of the Finno-Ugric languages and any Samoyedic tongue would be even greater.

THE DIFFUSION OF LANGUAGE

Languages have diffused spatially since their inception. The diffusion of some languages has come at the expense of many others. Ten thousand years ago the human race consisted of one million people speaking an estimated 15,000 languages. Today, with a population of 6 billion-plus, only 5,000 languages are recognized. Some experts feel that all but 300 languages will be extinct within the next 100 years. Each passing decade sees the extinction of more minor languages; for example, the Cornish language of Western England has less than a handful of speakers.

Different types of cultural diffusion have helped shape the linguistic map. Relocation diffusion has been extremely important, for languages spread when groups in whole or in part migrate from one area to another. Some individual tongues or entire language families are no longer spoken in the region where they originated.

One of the most impressive examples of the linguistic diffusion is that of the Austronesian languages. From a presumed hearth in the interior of Southeast Asia speakers of the language family first spread southward into the Malay Peninsula. Then, in a process lasting several thousand years, and requiring remarkable navigational skills, they migrated from the islands of Indonesia and sailed in small boats across the vast uncharted expanses of ocean to New Zealand and the islands of the South Pacific.

The most remarkable of all the diffusionary achievements was made by the Polynesian people, who sailed across the Southern Ocean in outrigger canoes against prevailing winds and ocean currents. These Polynesian sailors had no way of knowing ahead of time that land existed as no humans had previously found the Hawaiian Islands.

The relocation diffusion that produced the remarkable present distribution of the Polynesian languages has long been the subject of controversy. How, and by what means, could a traditional people have achieved the diffusion? Geolinguists Michael Levison, Gerard Ward and John Webb answered these questions by developing a computer model incorporating data on winds, ocean currents, vessel traits and capabilities, island visibility and duration of voyage. Both drift voyages, in which the boat simply floats with the winds and currents, and navigated voyages were considered.

After doing more than 100,000 voyage simulations, they concluded that the Polynesian triangle had probably been entered from the west, from the direction of the ancient Austronesian hearth area, in a process of "island hopping"—that is, migrating from one island to another one visible in the distance. The core of eastern Polynesia was probably reached in navigated voyages, but once attained; drift voyages easily explain much of the internal diffusion. According to Levison and his colleagues, a peripheral region, the "outer arc" from Hawaii through Easter Island to New Zealand, could be reached only by means of intentionally navigated voyages—truly astonishing and daring feats.

Indo-European Language Diffusion

In February 1786 Sir William Jones, a British judge stationed in India, recognized that Sanskrit bore a striking resemblance to two other ancient languages of his acquaintance, Latin and Greek. The Sanskrit word for father, transliterated from its exotic alphabet, emerged as *pitar*, astonishingly similar, he observed, to the Greek and Latin *pater*. The Sanskrit for mother was *matar*; in the Latin of his school days it was *mater*. Investigating further, he discovered dozens of similar correspondences. Though he was not the first to notice these similarities, no one before Sir William Jones had studied them systematically. The Sanskrit language shared with Greek and Latin "a stronger affinity . . . than could possibly have been produced by accident; so strong, indeed, that no philologer could examine them all three, without believing them to have sprung from some common source, which, perhaps, no longer exists." (Crum, et al., 1986)

Two centuries of linguistic research have only strengthened Jones's basic proposition. We now know that the languages of about one-third of the human race come from this Indo-European "common source." These include the European descendants of Latin, French and Spanish, a great Slavic language, Russian, the Celtic languages, Irish and Scots Gaelic, and the offshoots of German—Dutch and English. A second important breakthrough in the search for the truth about "the common source" came from the folklorist Jakob Grimm, better known, with his brother Wilhelm, as a collector of fairy tales. "Grimm's Law" established beyond question that the German *vater* (and English *father*) has the same root as the Sanskrit/Latin *pitar/pater*. Words such as *me, new, seven,* and *mother* were also found to share this common ancestry. With these revelations the Indo-European basis for the common source was clear.

It is sometimes said that you can deduce the history of a people from the words they use. Clever detective work among some 50 prehistoric vocabularies has now led to a reconstruction of the lifestyle of a vanished people, the first Indo-European tribes, the distant forbearers of contemporary Europe. From the words they used—words for winter and horse—it seems likely that the Indo-Europeans lived a half-settled, half-nomadic existence. They had domestic animals, oxen, pigs, and sheep, they worked leather and wove wool, plowed the land, and planted grain. They had an established social and family structure, and they worshipped gods who are the clear ancestors of Indian, Mediterranean and Celtic deities.

Who these people were and when exactly they lived is hotly disputed. According to myth, they lived in the 'fertile crescent' of Mesopotamia, but this theory was exploded by 19th-century archaeology. Today, there are some who argue for the Kurgan culture of the Russian steppes, others for the farming culture of the Danube valley, while a recent theory suggests they lived in the Anatolia region of southeast Turkey. The dates vary from 6000 BC to 4500 BC. The most widely accepted theory locates the environment of the Indo-Europeans in a cold, northern climate in which common words for *snow, beech, bee,* and *wolf* played an important role. Furthermore, none of these prehistoric languages had a word for the sea. From this, and from our knowledge of nature, it is clear that the Indo-Europeans must have lived somewhere in northern central Europe.

Two innovations contributed to the break-up of this Central European society: the horse and the wheel. Some of the Indo-Europeans began to travel east and, over the course of time, established the Indo-Iranian languages of the Caucasus, India, Pakistan and Assam. Others began to drift west toward the gentler climates of Europe. Their descendants are found in Greece, Italy, Germany and the Baltic. Both the Rhine and the Rhone are thought to take their names from the Indo-European word meaning *flow*. English has much in common with all these languages. A word like *brother* has an obvious family resemblance to its Indo-European cousins: *broeder* (Dutch), *Bruder* (German), *phrater* (Greek), *brat* (Russian), *brathair* (Irish), and *bhratar* (Sanskrit).

In later millennia the diffusion of certain Indo-European languages, in particular Latin, English and Russian, occurred in conjunction with the territorial spread of great political empires. In such cases of imperial conquest, relocation and expansion diffusion are mutually exclusive. Relocation diffusion occurred as a small number of conquering elite came to rule an area. The language of the conqueror implanted by relocation diffusion often gains wide acceptance through expansion diffusion. Typically, the conqueror's language spread hierarchically initially by the more important and influential persons and by the city dwellers. The diffusion of Latin with the Roman conquest, and Spanish with the conquest of Latin America, occurred in this manner. No more so in the world is the spread of the English language the result of conquest, both military and commercial, and relocation.

The Diffusion of the English Language

The spread of English is astonishing. About one-sixth of the world's population speaks English either as a mother tongue or as a second language. The language is scattered across every continent and is surpassed, in numbers, though not in distribution, only by the speakers of Chinese. Three-quarters of the world's emails and faxes are in English. So are more than half the world's technical and scientific periodicals: it is the language of technology from Silicon Valley to Shanghai. English is the medium for 80 percent of the information stored in the world's computers. Nearly half of all business deals in Europe are conducted in English. It is the language of sports: the official language of the Olympics. English is the official voice of the air, and the sea. Five of the largest broadcasting companies in the world (CBS, NBC, ABC, BBC, CBC) transmit in English to audiences that regularly exceed 100 million viewers.

English has a few rivals, but no equals. Neither Spanish nor Arabic, both international languages have this global sway. Although as we mentioned earlier, languages are diffused by conquest, some languages become so strong that they overcome conquest, survive and above all become influential. English—the lingua franca of the world, the language of business, of science and of the Internet—is one such language that has survived conquest and almost extinction.

The rise of English is a remarkable success story. When Julius Caesar landed in Britain 2,000 years ago, English did not exist. Five hundred years later, *Englisc,* incomprehensible to modern ears, was probably spoken by about as few people as currently speak Cherokee—and with about as little influence. Nearly a thousand years later, at the end of the 16th century when William Shakespeare was in his prime, English was the native speech of between five and seven million Englishmen.

Four hundred years later, the speakers of English—including Scots, Irish, Welsh, American and many more—travelled into every corner of the world, carrying their language and culture with them. Today, English is used by at least 1 billion people, and barely one-third of those speak it as a mother tongue. English is more widely scattered, more widely spoken and written, than any other language has ever been. It has become *the* language of the planet, the first truly global language.

The origins of the English language seem somewhat incredulous—the Frisian islands of the western coast of present day Netherlands. The language spread by conquest when, in the 5th century of the modern era, the Germanic tribes of Saxons, Angles, Jutes and Frisians invaded the Southern and Eastern parts of Britain. These peoples colonized the land and developed the language throughout England. Most of the new masters of the island spoke a similar language, Anglo-Saxon, later to be called Old English.

In the 9th century changes to the language were taking place as peoples from northern Europe, the Norse or Vikings, first came to the land as plundering raiders and later as conquerors. These peoples tried to spread their influence throughout England but were repulsed to the northeastern parts known as the Danelaw. Where the two peoples met, mingled, and married became the place where the new language of the country developed. The Hearth of the English language is not in the south around London or Winchester where the King lived, but in the northern parts where the Norse and the Anglo-Saxon intermingled. The countryside that is now Yorkshire is the true hearth of the language. It is from here the language develops and becomes strong and resistant to further conquest.

This third conquest occurred in the 11th century with the Norman invasion. French was the language of the conqueror. All major factions in the towns spoke French: the court, the law, and the clergy. Yet instead of the common people using the new language, English continued to dominate in the countryside. So that instead of the conqueror's language dominating, the language of the subjugated prevailed. It is in this period that English seems to have died. There are few written records in English. The government, religion and trade continued using either Norman French or Latin. Elite groups such as the Barons, churchmen, lawyers, and the like continued to use French; while in the countryside, the continuity of the English language in the mouths of the mass of ordinary people was never in doubt.

Why did English survive? Why was it not absorbed into the dominant Norman tongue? There are three reasons. First, the Old English vernacular, both written and spoken, was simply too well established, too vigorous, and, thanks to its fusion with the Scandinavian languages, too hardy to be obliterated. It is one thing for the written record to become Latin and French (writing was the skilled monopoly of church-educated clerks), but it would have needed many centuries of French rule to eradicate it as the popular speech of ordinary people. The English speakers had an overwhelming demographic advantage. Pragmatically, it is obvious that the English were not going to stop speaking English because they had been conquered by a foreigner.

Second, English survived because almost immediately the Normans began to intermarry with those they had conquered. Barely 100 years after the invasion, a chronicler wrote that "the two nations have become so mixed that it is scarcely possible today, speaking of free men, to tell who is English and who is of Norman race." Thirdly, and perhaps most importantly, in 1204, the Anglo-Normans lost control of their French territory across the Channel. Many of the Norman nobility, who had held lands in both countries and divided their time between them, were forced to declare allegiance either to France or England (McCrum, Cran and MacNeil, 1986).

With the rise of England as a World power in the first Elizabethan age and later the Stuart period, the diffusion of the language worldwide began. The first stage of the diffusion process was with the colonization of North America and the Caribbean followed a century later to Australia, New Zealand, and southern Asia, and later in the 19th century to many parts of Africa. As English diffused throughout the colonies it developed local characteristics. So much so that English spoken by an American differs strongly from that of an Australian speaker or a South Asian. Just as there are dialectic differences in Great Britain so too are there these differences throughout the English speaking world.

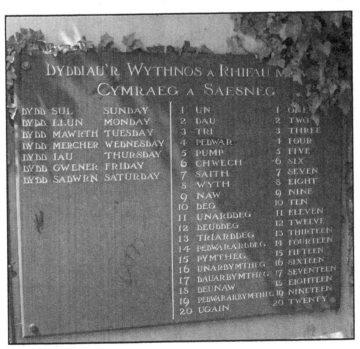

Figure 6.2   Welsh with English Translation in Conway, Wales. Both are Members of the Indo-European Family.

(© 2011, Leonard Peacefull)

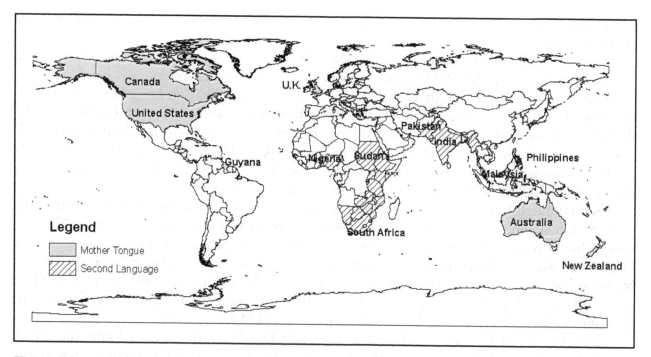

Figure 6.3   Countries where English Dominates as Either a First or Second Language.
(© 2011, Leonard Peacefull)

The emergence of English as a global phenomenon—as either a first, second or foreign language—has recently inspired the idea that we should talk not of English, but of many forms of English (Englishes), especially in Third World countries where the use of English is no longer part of the colonial legacy, but the result of decisions made since independence. But what kind of English is it? This is a new and hotly contested debate. The future is unpredictable but one thing is certain, the present flux of English as a multinational standard and international Babel is an ongoing process.

Frenglish! (Use of Borrowed English Words in the French Vocabulary)

In the 1980s a decision by the Académie Française to restrict the French people from using English words that had invaded the French vocabulary such as: *l'weekend* and referring to the music player of the time as a *balladeer* instead of a walkman, or saying "Nous irons de Paris à Londres en *gros-porteur*" instead of "en *jumbo-jet*" brought about different reactions from English and American commentators. These commentators highlight the fact that the French language has insinuated its way into English vocabulary; expressions such as: "Bon voyage," "Faux-pas," "laisser-faire," "esprit de corps," "a propos," and "vinaigrette." Not only are these words used without complaint in English speaking world, they are usually correctly pronounced, indicating an embrace of the French culture. One amusing commentary was written by Ellen Goodman of the Boston Globe who wrote:

> "Here is a true cause célèbre. Who made us wear lingerie, turned cooks into chefs, dances into ballets, and bunches of flowers into a bouquet? If they make it de rigueur to eliminate Americanisms, we shall refuse to eat our apple-pie a la mode and our soup du jour."

(For more of this see Goodman 1983)

Language and Culture in the Future

From the forgoing we can see that language is an import cultural tool, it has the ability to define a social group. On an individual level, language is an important way to establish ones unique identity. Each person has a unique set of linguistic resources on which to draw and each person makes unique, creative uses of these resources (Johnstone, 2000).

When people, using a common language, communicate with one another they do so by relating to each other with commonly understood words that translate ideas. The cultural difference is in the way that individuals transform these ideas into words, sentences and phrases, that have meaning. Language in any form is a natural extension of an individual's desire to express their self. Respect for cultural identities seems to be of paramount importance as global connections become stronger and more widespread.

As the world adapts to escalating levels of globalization, it is important for individual groups to maintain their cultural identity, while continuing to assimilate new cultural identities. However, as with most things global there is concern that many cultures and their languages will surrender to the onslaught of globalization. For example, in Switzerland, a polyglot country, there is concern that the four language-based culture groups, French, German, Italian, and Romansh, are in danger of losing their identity as English permeates the nation as the lingua franca. If the Swiss abandon their multilingual culture what does the future hold for other languages? Why should a German and a Bulgarian learn each other's language when learning English is more beneficial? It has been argued that future generations may forsake their cultural language and learn a lingua franca instead such as English or maybe Chinese. If this is so, what happens as the barriers between cultures are destroyed and cultural identities merge into one larger entity. Perhaps the solution lies in the preservation of language people will protect their cultures from extinction.

BODY LANGUAGE

Body language, or **kinesics,** is communication through gesture, expression, and posture. This form of language can be conveyed both consciously and unconsciously by an individual during the communication process. Due to this, body language makes it difficult for individuals to hide their emotions during face to face confrontations. By studying kinesics, much can be learned about an individual without actually conversing.

Body language can be looked at as the most important part of the communication process. In addition to the spoken word and tone of voice, scholars have suggested that 55 percent of what is communicated through conversation with others is communicated through body language. This section will focus on how body language is perceived across different cultures. Many variations of body language exist across different cultures. While some gestures are universal, others can be perceived differently across different cultural groups. What is seen as a permissible gesture to one group may be viewed as obscene by another.

The smile is the most universal form of **nonverbal communication.** Almost all cultures accept it as a symbol for happiness, amusement, and friendliness. There are some slight differences however, in how the smile is used worldwide. For example, while walking the streets of a North American city, smiling at a complete stranger is generally seen as a sign of politeness. If one were to do the same in Korea, the recipient of the smile may find it to be an inappropriate gesture because in Korea, a smile usually indicates embarrassment, not pleasure. In general, people of Asian cultures smile less than westerners.

Other than smiling, there are many other forms of body language used in the greeting process. For western cultures, it is most common to shake someone's hand firmly while engaging in intermittent eye contact. Other cultures may put more emphasis on physical contact. Places such as Cuba, Portugal, Spain, Italy, the Middle East and Eastern Europe typically greet by exchanging kisses on the cheek. Eastern cultures, however, have a much more formal approach to the greeting process. For example, most Asian cultures greet each other by bowing with their hands at their sides. The depth of the bow is related to the amount of respect due to that person. Also, in Japan, prolonged eye contact is seen as an invasion of privacy.

Body proximity is another important form of body language. Most cultures have differing views as to what constitutes proper personal space. In North America, the amount of personal space generally required is about the length of an arm. Other cultures are more comfortable with a closer proximity. People of France, the Middle East and Latin America often communicate within very close proximity to one another while also engaging in prolonged eye contact. Engaging in prolonged eye contact while in close physical proximity indicates active engagement and an interest in communication with others. Finally, in some areas of the world, personal space is determined based on age, gender, social status, or a variety of other factors.

Insult signals are perhaps the most varied form of body language across cultures. These signals are derogatory gestures that humans direct at one another to show their displeasure. There is no other species that can match the human range and diversity of offensive signals. When studying cultures, some insult signals are universal, while others require special local knowledge in order to understand their meaning. Universal derogatory signals can be used to insult an individual in a number of ways. Based on their origin in human infancy, these gestures are understood across most, if not all cultures.

The most common of these derogatory gestures is laughter. Laughter finds its origins in infancy as infants become startled by their parents during play. Examples include a mother tickling or swinging a child through the air. When this occurs, infants laugh as a sign of relief. Laughter indicates understanding that the action was strange or unexpected but not alarming because the action was initiated by the parent. When humans laugh at one another the implication is that the other person is perceived as somewhat strange yet nothing to be worried about. This action can be belittling to an individual and is why laughter often provokes an aggressive response.

Another universally derogatory signal is sticking out one's tongue. This form of insult is known as a rejection signal. It originates from an infant's rejection of a food, especially from the breast or bottle. As infants are unable to speak, this signal of rejection is developed early in life.

Humans have also developed universal displays of superiority. These displays show dominance of one individual over another. Examples of this are holding one's head high or wearing a permanent sneer when engaged in conversation. These are status displays that attempt to belittle other people. Superiority insults are a type of animal instinct that all humans possess. In the wild, animals exhibit their dominance by standing tall and looming over others within the group. Often, the animal that stands taller and shows less fear will win the standoff and establish authority. Humans simply mimic this dominant behavior in their own environments.

There are literally hundreds of thousands of localized insult signals. These are signals that are not understood universally, and often require familiarity with the cultural group using them in order to clearly understand their meaning. Listed below are a few examples:

- In certain South American countries, cupping one's hands just below the chin signals stupidity. It is intended to duplicate an imaginary goiter, which is itself a signal for idiocy.
- In certain areas of Spain, people rest their head on an imaginary hand as an insult. This imaginary hand supposedly represents a mother's hand, which signifies immaturity.

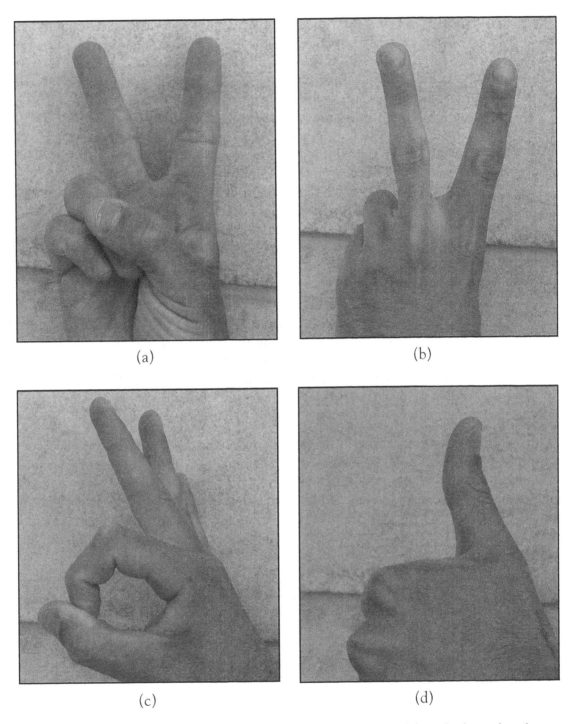

(a)   (b)

(c)   (d)

Figure 6.4   These four pictures show familiar hand gestures (a) with the palm facing is the sign for peace, (b) with the knuckles facing, in Britain is an obscene gesture; (c) may be ok in some places however, it is another obscene gesture in parts of Europe; while in Japan it is a sign for money; (d) again this may mean ok or a sign used by hitchhikers but again in parts of Southern Europe you could be in trouble using it as it has a similar meaning to the American single digit. (© 2011, Leonard Peacefull)

- In several European countries, a popular signal is to hold out one's hand, palm up, in front of your chest. This represents how long a beard can grow while listening to the boring speech of the person being insulted.
- People in France, Italy, Spain and Greece use the phallic symbol of a forearm jerk as an insult. This emulates an erect penis being formed by a clenched fist and a raised arm. This action is intended to show dominance over another person.
- While many countries use the thumbs-up signal as a way of saying "all is well," people of Sardinia and other areas of the Mediterranean use it to say "sit on this." This has created a great deal of confusion for foreign hitchhikers.
- The two-fingered V is a sign that has also created much confusion across cultures. Most countries recognize it as the victory, or peace sign, which was coined by Winston Churchill during World War II. In England, however, it has two different forms. If your palms are facing outward when making the V, it is the sign for victory. If your palms are facing inward, however, it turns into the worst obscenity of the English culture. This practice dates from the Anglo-French Hundred Years War when British long bowmen, if captured by the French, would have the two fingers of the right hand cut off. At the battle of Agincourt, to show despise of their enemy, all the British bowmen stood with the two fingers of their right hand raised in mockery toward the French lines.
- Different cultures also have different variations of the phallic middle finger gesture. Most countries point the middle finger straight up in the air with their palm facing toward them. In Arab countries, however, the gesture is performed by holding your hand out with your palm facing down and bending your middle finger.

As the world becomes increasingly integrated and globalized, it is important to become aware of cultural differences. Understanding another culture's practices, beliefs, and mannerisms is critical in order to dismantle cultural barriers. Having an appreciation of these differences in body language and gestures will save a great deal of embarrassment, misunderstanding, and antagonism in the communication process, and avoid possible trouble for the unwitting individual. In effect, creating a more unified and harmonious world built on understanding and effective communication.

KEY TERMS

Dialect	Polyglot	Language diffusion
Pidgin Language	Bilingual	Kinesics
Creole Language	Language families	Non-Verbal Communication
Lingua Franca		

DISCUSSION TOPICS

1. Language is the essential cultural trait; it distinguishes who we are as individuals and members of cultural groups.
2. Why is it important to be aware of differences in body language when visiting other cultures?

CHAPTER 7

The Diversity of Religion and Belief

If studying cultural variations were simply a matter of language differences then cultural understanding would be easy. However, cutting right across the cultural spectrum and language barriers is a system of beliefs that tend to complicate our appreciation of cultures. An awareness of the geography of religion, its origins and development, reveals important insights about the significance of religious beliefs and teaches us to understand more about the world's various cultures. Why is it that different cultures with diverse cultural traits speaking unrelated languages practice similar religions? For example, Albanians and Indonesians may hold similar views on Islam but their cultures are distinctive; while congregations in the Democratic Republic of the Congo and Connecticut share the same Catholic creed but do not speak each other's language. In each case the cultures of both groups are very diverse from the other.

Each of the world's main religions has a distinctive geography. Christianity's more than one billion adherents are located largely in Europe and the Near East, the Americas and Australasia. Islam has diffused from its birthplace in western Arabia through the northern half of Africa, central Asia and India and into Indonesia. Hinduism and Buddhism are highly localized; the former being largely confined to the Indian peninsula and the latter to East Asia.

It is also possible to divide the major religions into various subgroups. For example, if we examine the Christian subgroups within the United States, we find strong zonal pattern. The Roman Catholics are strongly represented in New York and the industrial northeast; Baptists in the Southern states, the Lutherans in Wisconsin, Minnesota and the Dakotas; and Mormons in Utah. Even within metropolitan areas geographic differences in religion may occur with Christian Protestant churches being found most frequently in the wealthier suburbs.

THE GEOGRAPHIC IMPACT OF BELIEFS AND RELIGION

Throughout the world there are a bewildering variety of beliefs that may take on many forms. However, it is religious belief that plays a central part in defining our various cultures. There are many definitions of religion but for the purpose of this introduction we will concern ourselves with *specific fundamental beliefs adhered to by a group or sect*. Therefore religion bestows a culture with a system of interlinked values that enter into several aspects of human geography.

One clear geographical aspect of a belief system is where it gives special value to particular locations. These range from prehistoric monuments (such as Stonehenge) to former hunting lands such as the homelands of some Aborigine tribes in Australia. Other geographic aspects may not be as well defined, or as clear to observers in the west, as the restrictions applied to agricultural practices. Many religious beliefs impact the development of agriculture through constraints on diet and the symbolic significance given to animal life. This can lead in many cases to major economic and health problems for the communities in which these practices take place.

To illustrate this take a look at the dietary restrictions placed on the adherents of many of the major religions. Although humans have developed as omnivores, a large proportion of the world's population restricts its diet in some degree. For example, the 300 million Buddhists are generally vegetarians, while Hindus (about 800 million) may not eat beef and the 20 million Jews may not eat pork. Smaller groups may have still more precise rules; for instance, the Jain communities of India (approximately 3 million people) are forbidden to kill or injure *any* kind of living creature.

Consequently, cattle throughout most of India are used only as draft animals and for milk production. Under the Hindu doctrine of *ahimsa*, the slaughter of cows is prohibited in many of the Indian states, so that old and unproductive cattle increase pressure on grazing resources. It has been estimated that the number of such "surplus" cattle is between one-third and one-half of the total cattle population.

An even more extreme view of the importance of cattle is taken by the herding cultures of eastern and southern Africa. Among groups such as the Masai, the number of cattle a man has is directly related to his prestige and wealth and can serve as the means of exchange, most notably in the purchase of brides. The attitude toward cattle among these herding tribes has a religious component (cattle are said to be "the gods with the wet nose"), although the emphasis appears to have more to do with their convenience as a means of exchange. In this instance the numbers rather than the quality are important, which has an adverse effect on the standards of livestock.

Another significant religious attachment to animals is that of the Muslim view of the pig—an unclean animal. Consequently, in Malaysia, which has both Muslim and non-Muslim communities, pigs are reared for food only in the Chinese enclaves while the native Malay population follows the Islamic code and deprives itself of an important source of food.

These variations are important in determining the cultural mosaic of the world. But where did it all begin—if there ever was a true beginning?

In the Beginning

The dawn of religion, tens of thousands of years ago, accompanied many other significant developments: the making of tools, clothing and ornaments; the construction of shelters, the control of fire; the beginning of a symbol system of writing that allowed the development of language; and the imaginative projections that resulted in art and religion. While we may never fully understand the inspiration that created them, graves and paintings from the Paleolithic era hint at practices and rituals that might be called religious.

As cultures settled into a particular pattern, as agriculture activity led to more sedentary lifestyles, religions became more formalized. For the hunter, religion could be individualistic and his worship was concerned with personal health and safety. The collective concerns of farmers were based on the calendar: the cycle of rainfall, the seasons of planting and harvesting, the rise and fall of waters to irrigate the crops. Religions responsive to those concerns developed rituals appropriate to seasons of planting, irrigation, harvesting and thanksgiving. An established priesthood was required, one that stood not only as intermediary between people and the forces of nature but also as authenticator of the timing and structure of the needed rituals.

Throughout the natural world there are forces in nature that can evoke fear and awe, dependency and anxiety: thunder and lightning storms, earthquakes, volcanoes, the ocean crashing on the shore, wind, rain and fire. The landscape prompts human responses: mountains soaring high appear to touch the heavens; a cave dark and dank presents a hideaway filled with ominous foreboding; a major river coursing with powered destruction limits the possibilities of daily life. To early mankind, these forces contained some form of sacred presence, some supernatural force that required their attention and deference; consequently in every culture throughout the ages, humans have sensed a power greater than themselves. Surrounded by these phenomena, which they do not understand and by forces they are unable to control, they have perceived a sacred presence. Humanity responded to these supernatural forces with fear and respect. In attempting to understand or gain favor from these unknown forces they developed forms of worship, recognized certain phenomena as sacred and kept them apart from ordinary activities. They have understood that there is a power greater than themselves and have sought some form of salvation through requests for spiritual or divine intervention.

Some cultures recognized sacred spirits in every part of their surroundings. Some developed a belief in gods—superhuman beings who interested themselves in the affairs of humans. Some cultures worship multiple gods while others believed in one **Supreme Being.** In order to have their worship accepted by their chosen deity, humans had to determine certain rules on how to relate to the divine being; how to behave in order to gain protection in this life; how to be protected from evil and eventually to achieve salvation in an afterlife. To placate these needs they performed special ceremonies, dances, rituals, offering gifts and sacrifices to beseech the gods or spirits.

Mankind cannot control, only respond to, the climate, the weather and annual variations in light, heat and moisture. Certain cycles cannot be interrupted, reversed, or denied, for example, the seasons of the natural world. There too are cycles of human life, from birth to death with landmarks in between such as puberty, marriage, childbirth and the death of parents and elders. Every one of these powerful experiences in the human landscape has inspired religious beliefs and various forms of worship.

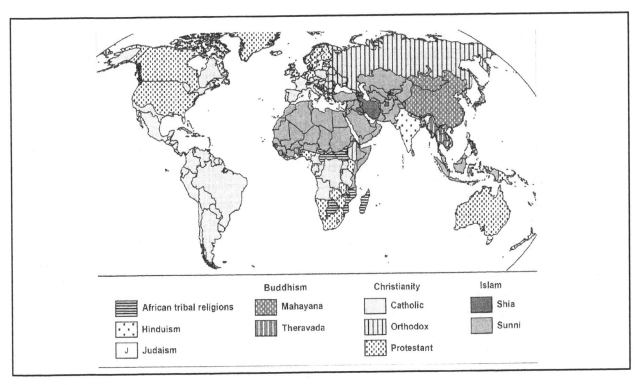

Figure 7.1   Major World Religions. (© 2011, Leonard Peacefull)

RELIGION AS A CULTURAL SYSTEM

Religion is an integrated complex of meanings, symbols and behaviors articulated by a community of adherents. Like other elements of culture, religion encompasses a set of concepts that govern the ways in which people understand, act within and influence the world that they inhabit. However, religion differs from other elements of culture in its particular concerns. A religion consists of a set of beliefs and practices whose truth is inferred by faith that ultimately relates to superhuman entities assumed to possess attributes or powers superior to those of ordinary mortals. These superhuman entities are most often represented as deities, but can take other forms as well, such as venerated ancestors, nature spirits, or persons who have achieved a state of spiritual perfection. Whatever form they take, these entities are considered by believers to exert crucial influences, directly or indirectly, for good or for ill, within and beyond the realm of human affairs. For believers, a religion offers a means of comprehending the nature of these entities as superhuman agents, their authority and effectiveness within the cosmos and the ways in which they interact with humanity or provide archetypal models for human behavior (Stump, 2008).

This definition of religion excludes vague or incidental notions of magic and superstition, as well as ideological systems that are not primarily concerned with matters of spiritual faith such as Marxism or nationalism. It asserts that religion as an identifiable element of culture differs explicitly in its concerns from other types of belief systems or models of reality. Thus, the definition of religion focuses on conceptions of the superhuman and avoids the inherent ambiguity of approaches that represent religion in purely subjective terms. Neither does the definition expand the concept to include any perspective that involves a sense of reverence, such as some forms of environmentalism. At the same time, this definition identifies significant points of correspondence among spiritual belief systems articulated by culture groups throughout the world. Such belief systems take diverse forms, yet the definition is sufficiently flexible to accommodate quite varied traditions and so remains useful in a comparative context.

The development of religions and beliefs is a consequence of mankind's consciousness and ability to ask questions that have led, through reason and observation, to query their own existence. The most articulated question throughout cultures is *who am I and why am I here?* To try to formulate some kind of explanation, cultures have developed myths and parables—narratives dramatizing the origin, destiny and interactions between humans, nature and the gods or spirits. Some stories provide answers about their "gods'" importance to the culture. Long before written language, these stories passed on the experiences, revelations, beliefs and promises of ancestors.

Another quest of early cultures was the need to gain control over the elements. Faced with cataclysmic forces, humans often addressed spirit beings in order to gain some sort of comfort in a world beyond their understanding or control. They explained the forces and phenomena of nature in stories and concepts that lesser members of the group could understand. Sometimes this understanding comes in the form of **rituals**—patterned activities with prescribed rules and outcomes, practices that allow people to face the unknown together. These practices include words (spoken or chanted), music, dance or processionals and a whole host of sensory stimuli to fully engage the participants. As these practices are taught from generation to generation, they form bonds in the present and connect to an earlier time beyond individual memory. Conceptually, the ritual ceremonies repeat and reinforce the beliefs that unite a people.

For a community or culture to unite people within it, it must share its rules of morality. Even before the organization of states and nations, religions provided laws of right and wrong and the boundaries of behavior within the community. Two questions are often most asked: (a) *What can I do to improve this life for myself and others?* And (b) *Does it really matter how I behave?* Myths and legends have served to answer these questions and to define the behavior whether narrating ideals or portraying final rewards and punishments, thereby expressing the concern of "the gods" for the group.

Following on from the above, we can see that religious faith is generally organized around two central concepts: a religion's **worldview** and its **ethos.** A worldview comprises a distinctive understanding of truth, reality and the forces that shape worldly events. In the context of religion, a worldview represents an integrated system of beliefs that specify the essential order of existence and the ultimate sources of causality, particularly as they relate to the nature and authority of the superhuman entity or entities recognized by the religion's adherents. As an expression of culture, a religion's worldview is socially produced and maintained, but adherents consider it to be absolute and incontrovertible.

The understandings of reality defined in the worldview thus have vital repercussions. These repercussions are clearly revealed in a religion's ethos or set of values, motivations and emotions which worshippers feel consistent with their worldview. An ethos essentially represents an understanding of how people should think and behave given the faith that they share. This sharing of the faith in their belief system underlies the importance of religion as an integral factor in culture.

Another truth or belief is the recognition that although death is humanity's ultimate worldly fate, we question what happens next. Some of the earliest proof of a religious sense of meaning beyond the human being's physical existence can be found in the care with which bodies tens of thousands of years ago were placed in graves. What comes after death is the unanswerable question, the experience all humans share yet can never know. Myths, ceremonies and intricate constellations of belief arise from looking deep into this mystery, central yet antithetical to human life and finding in it the inspiration for an entire system of beliefs and practices that declare human life, individually and collectively, to have meaning.

Religious Hearths

From the beginning to the present day there have been many hundreds of religions and beliefs. Some with huge followings have passed into history while others have stayed alive in small pockets throughout the world. Some faiths recognize a single deity (**monotheistic** religions) such as Judaism, Christianity and Islam, while others attribute divinity to a multitude of entities (**polytheistic** religions), such as Hinduism. Today we recognize five major world religions, Hinduism, Buddhism, Judaism, Christianity and Islam; along with their various sub-divisions.

Geographically it is significant that these five major religions have a definite spatial pattern with the monotheistic religions located mostly in the west and the polytheistic religions dominant in the east. Another distinguishing piece of the geographic puzzle is the role played by the Mediterranean Sea in separating Christianity to the north and Islam on its southern shores. (See the Map of World Religions, Figure 7.1)

We have already discussed the definition of a cultural hearth as the center of innovation from which cultural elements extend outward to exert influence on surrounding regions. Religious hearths, according to Stark and Bainbridge (1987), acquire their distinct identity in two stages. The first of these initiating events is the revelation where some occurrence or divine manifestation related to certain individuals takes place. For instance, the revelation of the Qur'an to Muhammad by the archangel Jibril is one such event. The second stage focuses on this revelation whereby the founders begin to develop a cohesive system, which they pass on to their followers. The continued vitality of the new religion and its diffusion from its place of origin depends on the acceptance by a critical mass of these followers. It becomes obvious that religious hearths will differ as their originating events differ.

A crucial distinction in the cultural hearth exists between the primary religion, those that support the initiating event and the secondary hearth where conflicting practices result in division—schisms, that fragment the primary religion, for example, the break up of Christianity into Roman Catholicism, Orthodoxy and Protestantism. While the primary tradition may coalesce and diffuse rapidly, sometimes within a generation—Buddhism, Christianity, for example,—others may take longer to develop. The development of secondary hearths often takes time. The new system may involve a revision of the worldview of the root tradition, with novel rituals, structures and sacred texts. In another context, the schism may lead to followers rejecting any modernizing process adhering to the more conservative forms of the past. Witness the break up of the Anabaptists that preceded the formation of the Amish and Mennonite churches.

Religious hearths have emerged in many places over historic time. On a global scale only a few of these hearth areas are of special significance. The map of religious hearth regions shows two areas that dominate in the development of world religions and can truly be referred to as religious hearths. These areas are referred to as the **Indic Hearth Region** and the **Semitic Hearth Region.** The Indic region is located in the Northern part of the Indian subcontinent—in the area that is today part of India, Pakistan and Nepal. The Semitic region is located in southwest Asia around the shores of the Mediterranean and Red Seas occupied today by Israel, the Palestinian State, Egypt and Saudi Arabia.

Although these hearths differ in their geography and the eventual religious manifestations, they developed along similar paths. The early religions in both areas, Vedism and Judaism, represent a consolidation of the traditions and belief of the cultures that inhabited these areas over a long period of time. This long process of consolidation and codification of the early cultures became widely incorporated in patterns of life within the hearth region so much so that it contributes to the sustainability over time, giving rise to enduring ethnic religions. Later religions that developed from these earlier forms within the same areas did so at a greater pace mostly in response to "revelations" of their founders. These later developments had a more profound universal impact, bringing about cultural and social changes that allowed their diffusion to occur rapidly beyond the hearth.

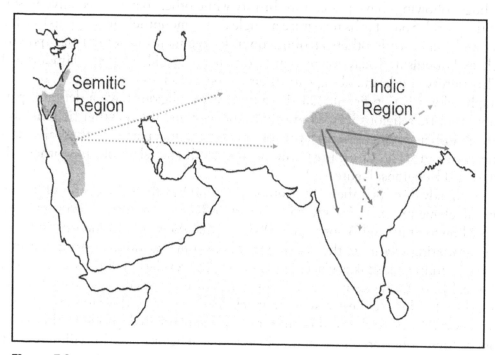

Figure 7.2   The Major Religious Hearth Regions. (© 2011, Leonard Peacefull)

The Indic Hearth

The Indic region is today part of India, Pakistan and Nepal, where the Himalayas rise majestically and the Ganges, Indus and Swat rivers drain and nurture the land and the people. It takes its name from the Indus Valley where between 3,000 and 5,000 years ago an ancient civilization flourished. Archaeological evidence suggests that these people's religion closely resembles elements found in modern Hinduism. Further evidence suggests that this culture may have lapsed or died, however around 3,500 years ago a new phase can be recognized. Aryan nomads began to move into the Indus Valley and settle. These Indo-Aryan settlements stretching across northern India established a cultural environment that not only gave rise to new cultural practices, such as the Sanskrit language but also the development of an enduring religion. This religion, **Vedism,** is based on the Sanskrit **Vedas** (books of knowledge); four books of hymns and mantras that are not only the core beliefs of this Indo-Aryan religion but continue to serve as the foundation of Hindu belief and to a lesser extent the other religions that originated in this region: Buddhism, Jainism and Sikhism.

The Semitic Hearth

As in the Indic hearth the origins of the enduring religions associated with this region can be traced to an early expression of the religious tradition, **Proto-Judaism.** These nomadic ancestors of Hebrew people or Israelites inhabited a region that is today occupied by Israel, the Palestinian State and Egypt. Their early wonderings are set down in the books of the bible Genesis and Exodus and in the Torah. Their worldview is centered on the belief in their covenant with a national god known as YHWH, with their ethos focused on divine law. These laws are codified in an early form of the Book of Deuteronomy. All these works together form the foundations of **Judaism** in the 6th century before the common era (BCE).

Like the religions of the Indic Hearth, the religions of the Semitic Hearth, endured and diffused across vast reaches of the globe, while giving foundation to other religions, Christianity and Islam. Unlike the religions of the Indic region, these latter two were able to develop and sustain in more literate cultures so that their religious codification, the core beliefs and practices became enshrined in scriptures and other texts. At the same time as Christianity and later Islam, were developing there occurred, in each of the regions of adoption, significant cultural processes that were to impact the peoples of the region and lead to the development of a critical mass of adherents. Consequently, these proselytic religions (where a person converts from one belief to another) were able to diffuse rapidly across the region and throughout the world.

THE IMPACT OF BELIEFS IN THE MODERN WORLD

Among the 7 billion people who inhabit the planet today, the Association Religious Data Archives (ARDA) recognizes 16 religious groups plus those who profess no religious affiliation.

Religions of most kinds lay heavy emphasis on continuity, tradition and strict adherence to long-established patterns of behavior. They have acted and continue to act, as a vital stabilizing influence, or—depending on one's point of view—an inhibiting drag on change.

Religion is often held to be a major factor inhibiting the spread of family planning practices. The moral values attached to the human fetus in Roman Catholic doctrine serve as a major barrier to the spread of abortion and certain contraceptive methods. This barrier may operate at the individual and family level for members of the Catholic faith or may become a matter of national policy in countries where there is a strong link between the Catholic Church and the state. Thus contraceptive devices are banned in Ireland and different attitudes are taken on abortion laws in the various states of the United States.

It is difficult to determine the importance of these attitudes from a strictly demographic viewpoint. Population-control practices are clearly described in the Old Testament and in Egyptian wall paintings dating from 5000 BC. There is ample evidence that human groups throughout history have been able to control family size when this was considered desirable. Attitudes toward what is the most desirable family size are demonstrably more important than which birth-control method is followed. Thus in Europe, a continent with lower birth rates than any area of comparable size in the world (generally about eight per thousand), there is no major difference between Catholic and non-Catholic populations at the national level. Countries in which contraceptives and birth control information are banned or restricted have birth rates just as low as those where both are freely available.

Rank	Religion	Adherents	% World Population
1	Christians	2,135,782,815	33.44%
2	Muslims	1,313,983,654	20.57%
3	Hindus	870,047,346	13.62%
4	Non-Religious	768,598,424	12.03%
5	Chinese Folk Religions	404,922,244	6.34%
6	Buddhists	378,809,103	5.93%
7	Tribal Religions, Shamanism, Animism	256,340,652	4.01%
8	Atheists	151,548,030	2.37%
9	Sikhs	25,373,879	0.40%
10	Jews	15,145,702	0.24%
11	Spriritists	13,030,538	0.20%
12	Bahai's	7,614,998	0.12%
13	Confucianists	6,470,714	0.10%
14	Jains	4,588,432	0.07%
15	Shintoists	2,789,098	0.04%
16	Taoists	2,733,859	0.04%
17	Zoroastrians	2,647,523	0.04%

TABLE 7.1   Religions of the World with the Most Adherents. (ARDA, 2011 and Religious Tolerance, 2011)

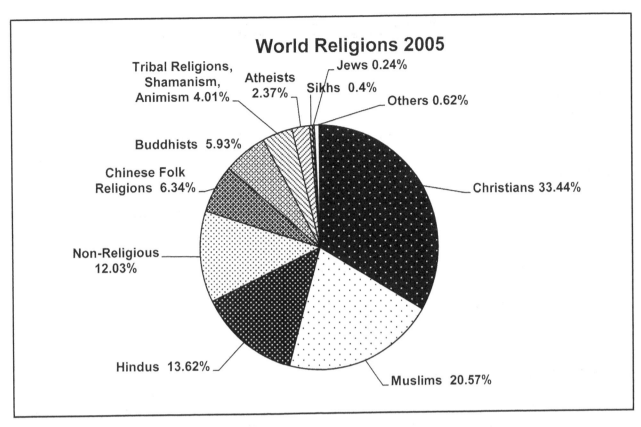

World Religions 2005

- Tribal Religions, Shamanism, Animism 4.01%
- Atheists 2.37%
- Sikhs 0.4%
- Jews 0.24%
- Others 0.62%
- Christians 33.44%
- Muslims 20.57%
- Hindus 13.62%
- Non-Religious 12.03%
- Chinese Folk Religions 6.34%
- Buddhists 5.93%

FIGURE 7.3   World Religions in 2005. (ARDA, 2011 and Religious Tolerance, 2011)

In other aspects of cultural traditions, the influence of religious beliefs on the acceptance of innovations is more specific and sometimes inhibiting. Take, for example, the 20,000 or so folk who follow the Amish religious faith. (For a more in-depth look at the Amish, see chapter 10). Located mostly in a tight region of the United States, Pennsylvania-Ohio-Indiana, in the midst of one of the more swiftly changing and highly modernized regions of the world, Amish communities stand out as islands of tradition and some might say cultural stagnation. They cling to their conservative traditions. Religious services are conducted in Pennsylvania Dutch. They wear traditional plain clothes. All forms of modern mechanization such as the telephone and electricity are shunned, while the horse and buggy continue to serve as the means of transport. Here, religious beliefs serve as the cement that continues to hold together a human group with a behavior pattern more reminiscent of 17th century rural Europe than contemporary North America.

Finally, a culture's place in the world—the Worldview—is often overlooked. In many cultures today, the system of belief conditions their view of the world and their place within it. In China today, the Worldview remains important in building construction. For example, when the new Chinese University of Hong Kong came to lay out its buildings at Sha Tin, an unusual locational factor had to be considered. A site was chosen in which the currents of the cosmos—*feng shui*—were harmonious. Chinese landscapes continue to have paths, structures and woodlots designed to blend with the natural landscape rather than dominate it. Even in the world's largest population the basic tenets of belief that go back many centuries still hold strong.

In the next section, we examine the development of some of these religions, those that play a major role in World cultures today.

THE MAJOR WORLD RELIGIONS

HINDUISM

Hinduism is truly an Indian phenomenon; it is practiced by 80 percent of the population. The religion is a way of life in India and forms an integral part of Indian culture. It is not easy to define Hinduism as a religion in the accepted western sense. Hinduism can best be defined as a way of life based on the teachings of ancient sages and scriptures.

The oldest of the five major world religions, Hinduism has evolved over the millennia to reflect and respond to the needs of the people at various times and in different landscapes and cultures. Through the centuries the Hindu tradition has proved capable of incorporating within itself many divergent beliefs and practices. It has always been a religion of many gods, many colors and many festivals. Within all this diversity Hinduism has been able to unify believers with each other within the world in which they live and everything around and beyond them; this is called Brahma, meaning the one.

The Origins of Hinduism

The word Hindu comes from the Persian word for river and is thought to refer to the people who lived in Indus (Shindu) river valley. The origins of Hinduism date back many centuries and have their roots in the ancient Aryan culture that inhabited the upper Indus River valley around 1,600 years before the Common Era (BCE). In fact, scholars have found evidence linking Hinduism with ancient civilizations that inhabited the region a thousand years earlier. These Indus cultures developed an environment that eventually gave rise to the Sanskrit language. This language provided a context for religious development in the form of a collection of hymns and mantras called the Vedas. The Vedas contain core beliefs that continue to serve as the ultimate foundation of Hinduism and other religions originating in this region.

The four Vedas, or books of knowledge, contain the core of the Indo-Aryan religion known as Vedism. The hymns and mantras contained in the Vedas refer to a pantheon of gods and outline the forms of worship and rituals required to appease and honor them. The *Rig-Veda* and *Sama-Veda* are the oldest and contain collections of hymns and mantras lauding the gods that date back to the middle of the second millennium around 1600 BCE. The *Yajur-Veda* is a manual for performing Vedism's key public ritual—sacrifice by fire. The fourth book called the *Atharva-Veda* includes mantras and incantations for health and good fortune and is used in domestic rituals. The interpretation of these books and rituals gave rise to a dominant priesthood known as *Brahmins*. With the formalization of rituals in a written language over time, Vedism came to be a highly structured religious system centered on a powerful priesthood.

By the early part of the first millennium BCE new thoughts and ideas emerged, which began to challenge the cultural trends of the Vedic beliefs. These cultural uprisings occurred in the eastern Ganges Valley, which has become a primary hearth for several new religious traditions and innovations. One of the earliest challenges to Vedic practices came from the ruling *Kshatriya* or warrior class. A series of writings called the *Upanishads* came to question certain Vedic practices. They focused on philosophical questions concerning the nature of reality and self and a reinterpretation of the concept of Brahma. Furthermore, they articulated a belief that a person experiences a continual cycle of reincarnation that in turn becomes the person's Karma. This law of karma (Sanskrit for action) asserts that the total effect of a person's actions and conduct during the successive phases of the person's existence have future reactions which we might regard as the person's destiny.

The Aryan Varnas or Caste System

When the Aryans moved into northwest India, they imposed a caste system to organize the new society created by their arrival. Initially there was a hierarchy of four varnas (castes), later expanded to include a fifth category. The caste system was designed to maintain a rigid social order. It set boundaries between the invaders and the indigenous peoples. Over the generations, the origins have been forgotten and the system became the foundations of the social order. There were four original varnas. The highest three were reserved for the conquering Aryans while the fourth consisted of the former inhabitants.

People are born into the caste of their parents. There is no mobility across caste lines during one's lifetime. Each varna is divided into a number of sub-castes, called a *jati*. Just as the varnas provide a hierarchy in society at large, the different jatis provide a social hierarchy within a varna.

This system of varnas and jatis serves two important functions. First, it assigns occupations. The varna and jati to which one belongs is usually identified with an occupation. As an example within the Vaishya Caste there are jatis of shopkeepers, farmers, workers in metal and wood, etc. Second, the system separates the members of the different varnas and jatis by a complex system of purity and impurity. The higher a person's varna or jati, the greater a level of purity the people must maintain, while it is felt that the lower ranks are more likely to transmit impurity. These purity restrictions appear most frequently in four areas: marriage, drink, food and touch. Marriage is possible only between members of closely related jatis. Touch is also an important factor in the social milieu; if for example, a shudra should accidentally brush against a Brahmin, this act would require the brahmin to undergo extensive rites of purification.

The higher Varnas, which excludes the Shudras, are designated as being twice born and are the only ones to be initiated in the Vedic Traditions. Twice born is a process that occurs when a caste member reaches maturity. The lower caste shudras on the other hand are excluded from this form of worship as they are not permitted to hear the Vedas read. This group, therefore, has their own rites of religion and their own priests.

In addition to the aforementioned castes is a fifth social group known as outcasts, or the Untouchables. This group is outside of the caste system so much so that the shudras cannot relate to them. They are assigned the worst occupations, such as street cleaners, garbage collectors, leather tanners, etc.

Varnas/ Caste	Occupation
Brahmin	Priests and religious officials
Kshatriya	Rulers and warriors
Vaishya	Farmers, merchants, traders and craftsmen
Shudra	Servants of upper castes and peasants

TABLE 7.2   The Four Aryan Varnas

HINDU BELIEFS AND DEITIES

Beliefs

Hinduism is not a homogeneous, organized system. Many Hindus are devoted followers of Shiva or Vishnu, whom they regard as the only true God, while others look inward to the divine Self (atman). But most recognize the existence of Brahman, the unifying principle and Supreme Reality behind all that is. Hinduism embraces a great diversity of beliefs unlike any other religion. A person can believe a wide variety of things about God, the universe and the path to liberation and still be considered a Hindu. These concepts have made Hinduism a more open-minded religion than many others.

While most Hindus worship local deities and characteristic beliefs, there are some beliefs held common to all forms of Hinduism. These fundamental beliefs include: the authority of the Vedas and the Brahmans (the priests); the acceptance of reincarnation; and the law of *karma* that determines one's destiny both in this life and the next. The ultimate goal of all Hindus is Moksha, the release from the cycle of Samsara (rebirth). There is not a specific belief in God or gods that is essential to the religion.

Hinduism's beliefs and worship are radically different from those of the monotheistic religions wherein God is omnipotent and everywhere throughout the universe. To many this goes against the evidences of human senses. The Hindu belief is quite simple: Brahman *is* creation; everything is an emanation from him. His essence lies in all created objects, including human beings. This means that the whole universe is actually one divine being. This simple notion has a stunning ramification; the soul of each person is Brahman.

The Hindu faith is rooted in the view that life is good; the human problem is how to enjoy life, or, more precisely, how to enjoy one's lives. To achieve this, Hinduism promotes four life goals: Dharma (virtue), Artha (success), Kama (pleasure) and Moksha (release).

- **Dharma** is the practice of virtue, the living of an ethical, correct life. The definition of what is virtuous, however, varies, depending on a person's caste. The primary virtue is to fulfill the duties assigned to one's caste. Hindus consider dharma the very foundation of life. It means "that which holds" the people of this world and the whole creation.
- **Artha** is the working for and achieving of success, in terms of both wealth and power. This means it is religiously important to be a successful businessman, to sell a lot of carpets, for instance, or to manage a successful restaurant. It also means that it is religiously good to serve on the city council, to be active in civic organizations, or even to become a politician.
- **Kama** is pleasure, usually understood as aesthetic pleasure of all kinds. This includes: the producing and enjoyment of art, music, dance, drama, literature, poetry and sex. (The "Kama Sutra," a well-known Hindu text, is about the aesthetic pleasure of men and women; it discusses beauty, music, dance and sexual activity.) *Karma* is a Sanskrit term meaning "action." It refers to a concept in which the results of one's actions accumulate over one's life. A good karma leads to a higher position in rebirth, while bad karma can lead to a lower position, possibly even one below the human race (like a goat or a bug). The more virtuous a person is in their present life, the higher they will be reborn in the next. And, of course, the higher one is born, the more enjoyable life will be. It is thus religiously praiseworthy to take part, to support, or just to appreciate any form of pleasure. This should always be done, of course, within the realm of dharma. A Hindu quote says Good Dharma equals good Karma.

- **Moksha** is striving for release from life (since, after all, it is bad). To achieve this, a person must turn their back on life and strive to live without the things that make up life. It requires a person to become an ascetic, a hermit and to spend time in contemplating the true nature of the cosmos and one's place within it.

The Deities

Western society in general associates Hinduism with a multitude of gods all having a need to be venerated and worshipped. True, the Gods and Goddesses of Hinduism amount to several thousand; however, each one is a representation of the many aspects of the one supreme Absolute known from the time of the Vedas as Brahman. It should be understood that although there are many manifestations of Brahman, each deity is really an aspect of the Brahman himself.

Consequently, there is not a specific belief in God or gods that is essential to the religion. Most Hindus are devout followers of one of the principle gods Shiva or Vishnu. Yet all Hindus accept the dominance of Brahman as the eternal origin who is the cause and foundation of all existence—the manifestations of a single Reality. Although Brahman is regarded as the Creator he is not worshipped in the same way as other gods because it is believed that his work—that of creation—has been done.

The main Gods of Hindu worship are Vishnu and Shiva.

- *Vishnu*, the preserver, is believed to be linked to a very early sun god and is considered by his worshippers to be the greatest among the gods. He is also referred to as Narayana. Vishnu preserves and protects the universe and has appeared on the earth through his avatars (incarnations), Krishna and Rama, to save humankind from natural disasters or from tyranny. Vishnu is represented in sculpture and painting in human form, often painted blue.
- *Shiva* is considered to be everything by those who worship him—he is a paradox: creator, preserver and destroyer. In Shiva opposites meet. Of necessity he is the destroyer because, without destruction, there can be no recreation. He is also the creator who is said to have danced across the earth creating life as he went. Consequently, he is the source of both good and evil who combines many contradictory elements. In pictures and sculptures, Shiva is represented as Lord of the Dance who controls the movement of the universe. He is also associated with fertility. Shiva has many consorts, including Kali, often portrayed as wild and violent, Parvati, renowned for her gentleness and Durga, a powerful goddess created from the combined forces of the anger of several gods. His city is Varanasi and any Hindu who dies there is believed to go straight to heaven.

Beside the main Gods are their consorts who are in their turn important Goddesses.

- *Lakshmi* is the consort of Vishnu. She is the goddess of wealth and good fortune and is therefore a popular goddess who is offered special worship during the Divali festival. In Hindu art, she is full-breasted, broad-hipped and is always smiling. She wears a red sari and coins rain down from two of her hands. In her other two hands she holds lotuses, representing the spiritual gifts she bestows. She is often shown seated on a lotus and being anointed by two elephants. (See also http://www.koausa.org/Gods/God6.html)
- *Parvati* is a consort of Shiva renown for her gentleness. She is depicted in Hindu art as a mature and beautiful woman, usually with Shiva.

- *Kali* is also supposed to have been a consort of Shiva. Her name means "She who is black." She is generally depicted half-naked, with a garland of skulls, a belt of severed limbs and waving scary-looking weapons with two of her 10 hands. She is often dancing on a prostrate Shiva, who looks up at her admiringly. Two of Kali's hands are empty and in the mudras (gestures) of protection and fearlessness. Her tongue is stuck out in order to swallow up evil and negative thoughts.

THE DIFFUSION OF HINDUISM

Hinduism developed from out of the Vedic tradition. Its hearth is the eastern Ganga (Ganges) River valley. A belief in reincarnation infuses the entire Hindu tradition and through Karma their deeds in the present life will shape the character of the next. This doctrine of reincarnation began to emerge as a counter to the ancient Vedic traditions. As trade and economic prosperity improved the lives of the peoples this cultural transformation diffused throughout the Ganga valley. Hinduism did not spring suddenly from the insights and revelations of a particular individual; neither did the early practitioners possess a group identity. Instead, there was a gradual acceptance by those who sought new meaning to the Vedic teachings. Consequently, Hinduism's diffusion was incremental throughout the Indian sub-continent not reactionary or revolutionary.

There is no one specific hearth for the religion but a series of locations across the sub-continent. Hinduism took shape from the combining of a variety of individually related local beliefs and practices. In fact, the local region deities or gods became merged with those of the formal Vedic tradition, which preserved a local articulation of belief within the greater Hindu framework. The local context was not only related to the deities but also to the varying practices within caste traditions. These spatial variations in the Hindu culture enabled the diffusion of religious innovations across the whole of the sub-continent. Thus, instead of local practices developing, a pan-Indian religion emerges.

The emergence of a continental religion is more surprising given the vastly different cultures that existed at the time. The northern part of India was culturally of Aryan descent. They were nomadic pastoralists whose language was either Sanskrit or another of the Indo-European family. The people in the south of India were part of the Dravidian culture that had existed long before the Aryan invasion of the north. Such is Hinduism that it exhibits a more diffuse ethnic character that it was able to be absorbed by the two differing cultures and to a lesser extent many minor culture groups as well. In the Gangetic hearth Hinduism was an amplification of the early Vedic traditions. While in the south, Hinduism emerged from the incorporation of the Indo-Aryan culture with that of the Dravidian. For example, the important Tamil gods became associated or identified with the northern Gods Vishnu and Shiva, as were the Tamil goddesses. Over time the two traditions became so merged that they were indistinguishable.

With the ascent of Buddhism and the later dominance of Islam, Hinduism was confined to the sub-continent and parts of Southeast Asia. However, in recent years with the migration of people worldwide the Hindu tradition can be found in the Southern Caribbean, East Africa, the Pacific Islands, Australia, North America and the United Kingdom.

BUDDHISM

Buddhism is more a way of life than a religion as people in the West may call their *monotheistic* traditional forms of worship. Buddhists have no god but a transcendental philosophy founded on the teachings of Siddhartha Gautama, the first Buddha. Buddhism is the second oldest of the five major religions evolving from within the Hindu Brahman traditions of the day, while very quickly developing in a distinctive direction. The Buddha not only rejected significant aspects of Hindu philosophy, but also challenged the authority of the priesthood, denied the validity of the Vedic scriptures and rejected the sacrificial cult based on these scriptures. Moreover, he opened his movement to members of all castes, denying that a person's spiritual worth was a matter of birth.

THE ORIGINS OF THE RELIGION

Buddhism is an off-shoot of Hinduism. Siddhartha Gautama, the founder of Buddhism, was for all intents and purposes Hindu. Siddhartha was born in a part of the Indian subcontinent that is now southern Nepal in 563 BCE. He was the son of Suddhodana, the ruler of the Skyas, a people inhabiting a country that lay on the border between modern Nepal and India. According to Buddhist teachings, at Siddhartha's birth the Brahmin Asita prophesized that he would either be a great ruler or a fully enlightened teacher. Siddhartha's mother died shortly after giving birth and his father brought him up in the lap of luxury with the training befitting a future king. Suddhodana wanted his son to become a monarch rather than a great religious teacher. Besides training in the martial arts, the king allowed his son to be instructed by Brahmans in the religion of the Vedas. The young prince then married a beautiful princess who gave him a son. Today we might say that his life was complete, but then a series of events lead to a dramatic change.

At the age of 29 Siddhartha ventured, for the first time, outside the palace walls and what he saw would change his life. He saw things he never knew existed. He encountered three signs, a decrepit old man, a very sick man and a corpse; sickness, old age and death had been kept from him, these three evidences struck him deeply and then he saw a sadhu (a Hindu religious recluse). The misery and brevity of human existence struck him with force and filled him with the desire to find a solution to the problems of life. These have become the famous four signs on which Buddha's teachings were to develop. So profound were the impact of these events on Siddhartha that he foreswore all worldly goods, shaved his head and assuming the name Gautma, he left his opulent life to wander through northern India as an ascetic, begging for food and living rough.

For six years, along with various other ascetics, he practiced different forms of meditation without finding the answers for which he was searching. He eventually left this group and continued his search alone. His search finally ended on the full moon day in the month of Veskha (April–May) in the year 528 BCE when he meditated under a Pippala (or Bo tree) a species of fig (*ficus religiosa*) near Uruvela on the Nairanjana river. The tree and the place together have come to be known as Bodh Gaya and is regarded as a symbol of the Buddha's enlightenment.

Gautma had become the **Buddha** (a term meaning the "Enlightened One"), by which title he was henceforth to be universally known. After considerable thought he decided to proclaim his discovery for the benefit of mankind. Determined to share his new insights of humanity with every man and woman, high and low caste, he set off to search for his five friends from his wandering days. At first he was hesitant about sharing the complexity of the system he had discovered with its opposition to the comfortable beliefs which then as now appeared to offer an easier solution to the spiritual needs of people. He found his friends in a deer-park that today is called Sarnath where he proclaimed the Four Noble Truths.

1. Life is suffering. This is more than a mere recognition of the presence of suffering in existence. It is a statement that, in its very nature, human existence is essentially painful from the moment of birth to the moment of death.
2. All suffering is caused by ignorance of the nature of reality and the craving, attachment and grasping that result from such ignorance.
3. By eliminating the cause, suffering can be ended.
4. There is a path by which one can end suffering; the Noble Eightfold Path, which consists of right views, right intention, right speech, right action, right livelihood, right effort, right-mindedness and right contemplation. These eight are usually divided into three categories that form the cornerstone of Buddhist faith: morality, wisdom and *samadhi*, or concentration.

The Buddha's message was simple: it followed the doctrine of Indian physicians of the age who, as many physicians do today, looked at a patient and followed a prescribed pathology—the symptoms, the cause, the prognosis of a cure and the remedy to achieve the cure. However, the principles that the Buddha espoused assumed a belief in reincarnation, the cycle of birth, death and rebirth—the wheel of life, which along with other matters of the world keeps mankind chained to the never-ending cycle of life.

The Noble Eightfold Path

The Buddha's path of practice is called the Noble Eightfold path. The eight components of this path, as presented in traditional order, could briefly be described as follows:

1. Right View (Understanding)—This is the right way of interpreting and viewing the world. It involves the realization of the three signata in all phenomena and of the Four Noble Truths as being applicable to the human condition. More generally, it involves the abandonment of all dogmatically held wrong views.
2. Right Intention (Thought)—The Buddha argued that all human thought and action spring from basic "intentions," "dispositions," or "roots," which are capable of deliberate cultivation, training and control. The three roots of wrong, unwholesome, or "unskillful" action are: Greed, Aversion and Delusion. The right intention, which the Buddhist path requires, is an intention which is free from these roots. The Buddha called the intention "that is free from greed and lust, free from ill will, free from cruelty."
3. Right Speech—Since speech is the most powerful means of communication, the Buddha emphasizes the cultivation of right modes of speech. These have been described as avoiding falsehood and adhering to the truth; abstaining from tale-bearing and instead promoting harmony; refraining from harsh language and cultivating gentle and courteous speech; avoiding vain, irresponsible and foolish talk and speaking in reasoned terms on subjects of value. Naturally right speech includes in the modern context right ways of communication whatever the medium used.
4. Right Action—This refers to willful acts done by a person, whether by body or mind. Under the former it involves such forms of ethical conduct as not killing (or harming) living beings, theft, sexual wrong-doing, etc. On the positive side, right action, also called wholesome deeds (kusalakamma), involves acts of loving-kindness (mett), compassion (karun), sympathetic joy (mudita), generosity (cga), etc.
5. Right Livelihood—This involves not choosing an occupation that brings suffering to others, e.g., trading in living beings (including humans), arms, drugs, poisons, etc.;

slaughtering, fishing, soldiering, soothsaying, trickery, usury, etc. This provides the economic blueprint for a truly Buddhist society.

6. Right Effort—This has been described as "the effort of avoiding or overcoming evil and unwholesome things and of developing and maintaining wholesome things" (yntiloka). Right effort enables an individual to cultivate the right frame of mind in order to accomplish the ethical requirements under right speech, right action and right livelihood. It is generally presented as a factor of mental training, enabling individuals to develop the sublime states of loving-kindness (mett), compassion (karun), sympathetic joy (mudita) and equanimity (upekkha). However, it has a general applicability and the effort could be directed to all wholesome activities.

7. Right Mindfulness—This is the basic Buddhist technique of cultivating awareness. The classic sutta on the subject is the satipahna sutta which will be considered briefly in the next chapter. Although viewed as a meditation component, in fact, right awareness has a wider applicability.

8. Right Concentration—This is the concentration of mind associated with wholesome consciousness which could be achieved through the systematic cultivation of meditation. Progress along this line is indicated by the achievement of the different levels of "absorption" (jhnas).

Of these eight components of the Path, the first two have usually been grouped under wisdom (pa), the next three under morality (sla) and the last three under mental development (bhvan). This classification is not quite satisfactory, but it does present a broad grouping that is useful in many contexts.

The first of these components (right view) is generally considered the most important, but there is no particular order of importance when it comes to the others. However, different traditions and exponents have put different degrees of stress on the different components. It will be seen that there is no single component of the path that can be called "meditation." However, in the course of time the component of mental development came to be regarded as meditation.

The five ascetics to whom he made this proclamation became the first members of the Buddhist Sangha (or Community of the faithful), which have survived to this day in unbroken succession. Together they agreed to preach the dharma, the doctrine of and the path to enlightenment.

Karma and Nirvana

Two of the central beliefs of Buddhism are the doctrine of karma and the attainment of Nirvana. Karma relates to a person's acts and their ethical consequences, which in turn leads to rebirth. Consequently, any good deeds are rewarded and evil deeds punished. One's karma determines such matters as one's species, beauty, intelligence, longevity, wealth and social status. According to the Buddha, karma of varying types can lead to rebirth as a human, an animal, a hungry ghost, a denizen of hell, or even one of the Hindu gods.

Buddhism does not actually deny the existence of the gods. It does, however, deny them any special role. The gods' lives in heaven may be long and pleasurable, but they suffer the same predicament as other creatures, being subject eventually to death and further rebirth in lower states of existence. Gods are not creators of the universe or in control of human destiny. Therefore Buddhism denies any form of prayer or sacrifice. Of the possible modes of rebirth, human existence is preferable because the deities are so engrossed in their own pleasures that they lose sight of the need for salvation.

The ultimate goal of Buddhism is to be released from the round of unique existence and suffering. To achieve this goal is to attain *nirvana*, an enlightened state in which greed, hatred and ignorance have been extinguished. Nirvana is a state of consciousness beyond definition where the enlightened individual continues to live, shedding any remaining bad karma until, at the moment of death; a state of final nirvana (*parinirvana*) is attained.

THE DIFFUSION OF BUDDHISM

The diffusion of Buddhism throughout much of the Indian subcontinent is credited to a very unlikely person, King Ashoka, originally known as Ashoka the Cruel, ruler of the Mauryan Empire that covered most of modern India and Pakistan. He had inherited the kingdom from his grandfather Chandragupta, the founder of the dynasty, by killing his brothers and so ensuring his succession to the throne. He also cruelly subjugated neighboring states to enlarge his Empire. His final cruel campaign was against the Kingdom Kalinga, known today as Orissa. Here he witnessed hundreds of thousands of men being slaughtered and he asked himself for what? These scenes of war led Ashoka to embrace Buddhism with a passion and from then on his conquests were by persuasion and conversion. Although he became a devout Buddhist, Ashoka was tolerant of all religions. He treated his subjects as equals regardless of their religion, politics and caste. Ashoka taught and persuaded his people to love and respect all living things. His laws limited hunting although it was permitted for food but the overwhelming majority of Indians chose by their own free will to become vegetarians. For the next 30 years Ashoka traveled across India spreading the Buddhist philosophy. His followers or missionary envoys traveled throughout Southeast Asia. Ashoka himself is said to have taken Buddhism to Sri Lanka.

After Ashoka's death, sometime during the 1st century before the Common Era, Buddhism underwent a split into two schools of thought with a distinct spatial variation: Mahayana Buddhism as a more radical view of the Buddha's teachings representing metaphysical ideas and lay devotions developed in Northern India; while Theravada, as a conservative interpretation of Buddhism flourished in Sri Lanka and southern India.

Trade helped the diffusion process as monks traveled from the hearth region into the Indus valley and Central Asia. There the Silk Road that traversed the lands north of the Himalayas and Tibetan plateau provided an important conduit for the spread of Buddhism. First, they followed the trade routes to the northwest into the kingdoms of Central Asia and later eastward into China. (See Map on the following page.) Because of its strong foothold in Northern India, Mahayanists tended to dominate these early missions. The flexibility in the teachings of the Mahayanists was a key factor in the acceptance of Buddhism by the nomadic peoples of Central Asia. They were able to adopt Buddhism without giving up the consumption of meat, a key element in their diet. Another important factor that assisted Mahayan monks was their ability to engage with all forms of society, especially in delivering medical aid. This adhered them to all levels of society and produced many converts. More importantly, the flexibility of Mahanyan philosophy allowed them to coexist with other religions and traditions of the region such as Taoism and Confucianism in China, the Bon in Tibet and Shintoism in Japan. As Buddhism spread through East Asia it died out as a religion in its hearth region of India.

There are many reasons why Buddhism may have declined in India. First, it is thought to have become absorbed in the changing Hindu traditions. Second was the lack of Royal patronage as the kings and Maharajas withdrew their support from the monasteries and universities. Third, the Brahmin priests saw Buddhism as a threat to their high caste status. Finally, the Muslim invasion in the early part of the second millennium devastated the Buddhist monasteries, as they were contrary to the teachings of the Qur'an. By the 13th century this destruction effectively eliminated Buddhism from the Muslim controlled Northern provinces.

One of the lasting strengths of Buddhism has been its ability to adapt to changing conditions and cultures. It is philosophically opposed to materialism, whether of the Western or the Marxist-Communist variety. Buddhism does not recognize a conflict between itself and modern science. On the contrary, it holds that the Buddha applied the experimental approach to questions of ultimate truth.

Although Buddhism in India largely died out, a small-scale resurgence was sparked by the conversion of millions of members of the untouchable caste. While under the Communist regimes in East and South East Asia, Buddhism has faced a more difficult time. In China, for example, it continues to exist, although under strict government regulation and supervision. Many monasteries and temples have been converted to schools, dispensaries and other public uses. Monks and nuns have been required to undertake employment in addition to their religious functions. In Tibet, the reigning government has tried to undermine the influence of the Buddhist monks.

Today Buddhism is a major part of the cultures in Tibet, Burma (Myanmar), Thailand, South East Asia, parts of China, Korea and Japan. Although its diffusion to the West was slow, Buddhism is undergoing a process of acculturation to its new environment. Although its influence is still small, apart from immigrant communities from East and South East Asia, new distinctively Western forms of Buddhism may eventually develop.

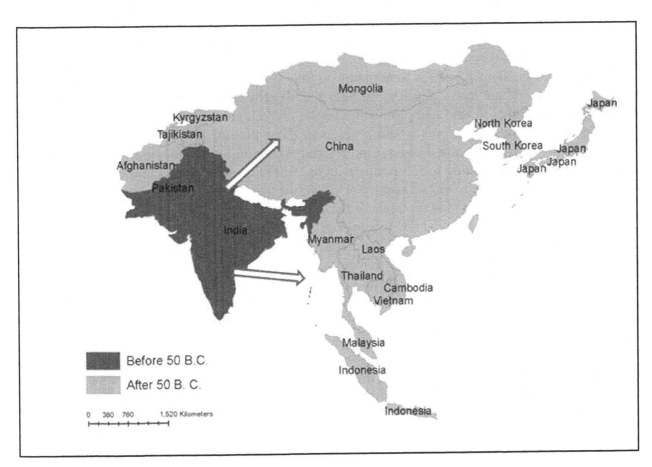

Figure 7.4   The Diffusion of Buddhism. (University of Akron produced)

JUDAISM

Elyssa Hilton, Temple Israel, Akron, Ohio

Judaism, a 4,000-year-old religion, is practiced by less than 1 percent of the world's population, yet it is influential in the foundation of the two most practiced religions, Christianity and Islam. From its inception, Judaism has brought a perspective different than other beliefs; monotheism, morality and rational. While it is a minority throughout the world, it can be found everywhere. Judaism came out of years of persecution and exile. The Jewish people consider themselves as all being part of the same family. The patriarchs of Judaism, Abraham, Isaac and Jacob, as well as the prophet Moses are revered by Christianity and Islam. Besides sharing the same foundation, these three religions all originate in the Middle East, the Semitic hearth. Before there were Jews there were the Israelites.

THE ORIGIN OF JUDAISM

The Israelites are defined as the chosen people of God, descendents of Jacob. The time of the Israelites is known as Proto-Judaism, before Judaism. This is the time before Moses led his people out of Egypt, the time of Noah, Abraham, Isaac and Jacob. Noah is said to be the first real man in Jewish history with a story foreshadowing important elements of the religion (Johnson, 1987). Both Noah and his descendent Abraham were men who followed God's word; Noah in his construction of the ark and Abraham in his journey and later the aborted sacrifice of his son, Isaac. Abraham believed the word of God who had promised a land on which could be built a great nation where Abraham and all his descendents would be blessed. That promise led Abraham from Haran to Canaan. In the city that is now Nablus, God assured Abraham he was in the Promised Land. At this time many were migrating for economic reasons, Abraham was traveling for religious reasons. He believed in God as the ruler and creator of all, yet he also believed other groups of people had their own God. His belief and acceptance of the Promised Land are the reasons why Abraham is considered the founder of the Hebrew religious culture in which two of its characteristics are the covenant with God and the donation of The Land. As a reward for his belief and acceptance of the Promised Land, Abraham was blessed with many sons.

One of Abraham's sons named Isaac and his son Jacob traveled the land and at various times, both had encounters with God. During one of Jacob's divine encounters he was renamed Israel, meaning God has striven, God has saved. After a life filled with travels and conflicts and to escape from famine Jacob, now named Israel, took his family to Egypt. While in Egypt he and his descendants, who are referred to as the Israelites, did not have good fortune. The Egyptian Pharaoh wanted to build the city of his dreams; therefore, he enslaved the Israelites as laborers. A condition these people endured until they were led to their freedom by Moses.

While Abraham, Isaac and Jacob are considered the patriarchs of Judaism, Moses is considered to be the fulcrum-figure, or hinge, around which Jewish history turns (Johnson, 1987). Moses, who was to become a great leader and prophet, was born to a slave. His mother wanted better for him and set him in a basket and placed it in wetlands. Here, Moses was found by the pharaoh's daughter who adopted him and raised him as royalty. However, many of his behaviors countered that of Egyptian royalty showing a kinship to the ethical principles and qualities of the Israelites. One such action was the killing and burying of an Egyptian after he beat an Israelite slave. To avoid punishment, he fled to the Arabian Peninsula where he lived as a shepherd until he was directed by God to return to Egypt to free the Israelites. This was the famed incident of the burning bush.

Once back in Egypt, Moses approached the Pharaoh to free the Israelites. When he would not, God is said to have inflicted the Egyptians with various punishments, the 10 plagues. After the tenth plague, which was the death of all first-born Egyptians, Moses led the Israelites out of Egypt to begin their exodus back to the Promised Land. These events mark the beginning of Judaism and are honored with the celebration and observance of the holiday Passover.

During the wanderings through the Sinai desert, Moses was given the laws or commandments of Judaism by God while he was on Mount Sinai. Of these laws and commandments, the Ten Commandments are the most well-known and the foundation for Judaism, Christianity and Islam, the three religions descending from Abraham (Hitchcock and Esposito, 2004). The first commandment is what moved Judaism from other religions of the times, belief in one God. Other commandments forbid idol worship, adultery, murder, theft and lying, while requiring the honoring of parents and observance of the day of rest, the **Sabbath.** These are only the beginning; there were actually 613 commandments handed down through Moses. These rational laws, the Mosaic Law, provide a moral code and a social code for the Jewish people. Where most laws of the time valued property above all else, these value human life; if man is created in God's image, then human life is sacred. Crimes against human life are capital, whereas crimes against property are not. By accepting these commandments, Moses made a covenant with God as did Abraham and Jacob before him.

These are the stories of Jewish history. They consist of the books of Moses; Genesis, Exodus, Leviticus, Numbers and Deuteronomy and constitute the holy book called the Torah. This written history makes the Jews the only people in the world who possess a historical record which allows them to trace their origins to remote times (Johnson, 1987). The Torah is not only a historical record; it lays out the laws and commandments of Judaism. Therefore, the study of Torah is considered a commandment equal to all others.

How This Makes Jews Different

With the commandments, the Jewish people set a moral code, making them the first people to say "Thou shalt not." The presence of this code and these commandments makes it not only illegal to kill, commit adultery, or steal, but also morally wrong (Kushner, 1993). With these laws combined with the monotheistic nature of the religion, the early Jews were different from other religions and cultures. Judaism is a way of life, not just a religion. Throughout history this difference from other cultures has led to serial persecution. Much like having to leave Egypt, the Jews have traveled to many parts of the world in order to follow their faith. In 586 BCE, the Babylonians invaded Jerusalem and destroyed the first Jewish Temple causing the Jews to be exiled to Babylonia. The Persians later conquered Mesopotamia, freeing the Jews and allowing them to return to Judea where they built a second Temple in the city of Jerusalem. During the Roman's occupation of their land, this second temple was partially destroyed (Bednarz, et al., 2005). Though this temple was never rebuilt, the western wall of it still remains. It is a place where Jews from all over the world go to pray. This western wall is sometimes called the Wailing Wall. It is after the destruction of the second temple that the diffusion of the religion takes place and Jews begin traveling and living in different parts of the world.

ACCULTURATION AND DIVISION

The diffusion of the Jewish people throughout Europe and North Africa led to their adopting certain aspects of the local culture, thus creating the "branches" known as Sephardic and Ashkenazic. While each of these groups still followed the Jewish doctrine, law and basic beliefs, the differences came about from cultural influences of where they were living. One example of this is in language. Ashkenazim, who had integrated to Northern and Central Europe, combined Hebrew with the German and Slavic languages to create Yiddish, while the Sephardim, who integrated to the Iberian Peninsula, combined Hebrew with Spanish to create Ladino (Stump, 2008).

Sephardic Judaism

Sephardim originated in Spain, Portugal, Western France and the Northwest tip of Italy. The word Sephardic is derived from the Hebrew word for Spain (Rich, 2010). During most of the period of Muslim rule of the Iberian Peninsula, the Jews were able to practice without persecution. It must be remembered that Islam as well as Christianity is based on the same biblical traditions as Judaism. Therefore, there is a Qur'anic acceptance of Jews and Christian as "People of the Book" (Stump, 2008). There was more cultural integration and less oppression in the lands where Sephardic Judaism developed. Much of their Jewish thought and culture were influenced by the Arabic and Greek philosophy and science (Rich, 2010). There became Jewish philosophers who wrote works that had an influence throughout the Iberian Peninsula and Mediterranean region, such as Maimonides of the 12th century.

When a less tolerant Muslim rule came into power, the Jews migrated toward the eastern Mediterranean area where the Muslim ruler was more tolerant. Some Jews stayed in Spain, but moved to areas of Christian dominance (Stump, 2008). However, circumstances were beginning to change; the Christians were coming into power. Out of change came the mystical expression of Sephardic Judaism named the **Kabbalah.** This is comprised of various writings including commentaries on the Torah that examine the nature of the divine, of creation and of evil. Consequences of prayer and obedience of divine law are also examined in this work. The approach is different than that of previous works of this nature (Stump, 2008).

Most of the early Jewish settlers in America were Sephardic. In fact, the first Jewish congregation in North America was in New York and started by Sephardim and is still active today. The first Jewish congregation in Philadelphia was also Sephardic and is also still active (Rich, 2010).

Ashkenazic Judaism

While the Sephardim were able to integrate within their communities, the Ashkenazim did not have this advantage. The word Ashkenazic comes from the Hebrew word for Germany (Rich, 2010). Ashkenazic Judaism originated in France, Germany and Eastern Europe and spread to a much larger area than Sephardic. These areas had Christian rulers who made it difficult for the Ashkenazim to integrate in the society. The Christians blamed the Jews for the death of Jesus. This was the basis for the animosity between the two religions. While the Jewish trade connections to the Middle East were valued, for the most part Jews and Christians were segregated in these areas (Stump, 2008).

Jews were excluded from many economic activities. However, the Christians had religious rules prohibiting the charging of interest; therefore, they were not involved in money lending the other.

This led to many Jews becoming bankers. While the Christian rulers were benefiting financially from the Jews and were cancelling debts owed to them, they resented the Jewish creditors (Stump, 2008). As a result of this animosity and resentment there were many acts of violence against the Ashkenazic Jews and they were expelled from many areas with laws prohibiting them from others. Thousands were killed by this violence and the Ashkenazic Jews spread to other parts of Europe (Stump, 2008).

The independent, inward looking nature and adherence to traditional Jewish law were reinforced for the Ashkenazim within this conflict and violence. Self-governing societies were created by the Jews which had enforcement of the regulations of Jewish Law. There was an emphasis on the study of Torah which created a strong religious scholarship as well as rabbinic academies which provided the final say on legal matters for the Ashkenazim (Stump, 2008). Out of these academies came rules such as the banning of polygamy and nonconsensual divorce. Rules like these established Ashkenazim as a distinct branch of Judaism. However, within Ashkenazim, unlike the Sephardim, there were disagreements which led to sub-divisions within Ashkenazic Judaism. These sub-divisions are Chasidim, Reform Judaism and Ultra-Orthodox.

Chasidic Judaism

Chasidic Judaism started in the Polish Ukraine. Its founder, Rabbi Israel ben Eliezer, is known to his followers as Ba'al Shem Tov, or Besht. The word Chasidic comes from the word chasid, meaning pious. The foundation for Chasidic Judaism is more mystical and spiritual than other forms. They believe that you can experience a communion with God not only through prayer, but in ordinary daily practices. The focus of Chasidic Jews is more on the emotional aspect of prayer rather than the academic, the experiencing rituals, rather than executing them. Hasidic Judaism spread throughout Poland, Lithuania, Ukraine and Hungary. Due to the differences in views from other Ashkenazim, they settled into their own communities and became secluded from the other Jews (Stump, 2008). Each of these communities is led by its own spiritual leader called a Rebbe. He is considered to be a righteous one. This is a position that is inherited. The Rebbe has the final say on all decisions in a Hasid's life (Rich, 2010). In the present day, Hasidic communities can be found throughout the world—Israel, New York, Boston and South Africa to name a few.

Reform Judaism

Reform Judaism originated in Germany. This movement was started in an attempt to modernize and liberalize Judaism (Stump, 2008). Supporters of Reform Judaism wanted to introduce secular disciplines into traditional Jewish education and to free the Jews from the traditional laws of exclusion imposed by European states (Stump, 2008). This is not a rejection of the basic beliefs of Judaism, but rather an attempt to adapt and practice those beliefs in more modern terms. With this movement of Judaism, the world began to see the Jews as a community rather than the chosen nation of God (Stump, 2008). Reform Judaism focuses on repairing the world, Tikkun olam. There is an embracing of diversity while at the same time asserting commonality (Union for Reform Judaism, 2008). Rather than focus on the strict Jewish laws, Reform Judaism turned the focus to morality, reason and justice (Stump, 2008). Reform Judaism has also spread throughout the world. In North America, there are three principles of Reform that separate it from the other practices. The first one is inclusion of all Jews, those who have chosen to be Jewish as well as the children of interfaith marriages that are being raised Jewish. Second, Reform Judaism practices equality of women in all parts of Jewish life. The last is acceptance of homosexuals (Union for Reform Judaism, 2008).

ULTRA-ORTHODOX JUDAISM

Ultra-Orthodox Judaism, which began in Hungary, was started by Ashkenazic Jews who remained committed to their traditional beliefs and practices. These Jews rejected modernity; the belief that the Torah expressly forbids deviation from established traditions was the thinking behind this movement (Stump, 2008). Some people refer to Orthodoxy as more observant. These Jews see themselves more as a continuation of traditional practices than a movement. They hold a dedication to the Torah both as oral and written law (jewishvirtuallibrary.org). The emergence of this came mostly out of rejection of the Reform movement, seeing it as a threat to the survival of Judaism (Stump, 2008). In many places, Ultra-Orthodox Judaism has been grouped with Chasidic. They hold to some of the same practices and beliefs; however, they do not separate themselves into their own community. While the Ashkenazim have split into distinct branches, they are all based on the same sources and beliefs. In North America, Conservative Judaism has emerged as a compromise of sorts between the Orthodox and Reform movements (Stump, 2008). In contrast the Sephardim have not split into factions.

Judaism is practiced all over the world. You can find synagogues of all types just about anywhere you may go. While there may be some differences in dietary laws, Sabbath practices or dress throughout the different branches, all Jews believe in one God. They also all hold the belief in the practice of the study of Torah and being a good, moral person. Judaism is more than a religion; it is a way of life, a culture. It is about overcoming challenges and hardships. Jews bond together in all circumstances and are more of a "people" than individuals. It was the first monotheistic religion and the beginning of others.

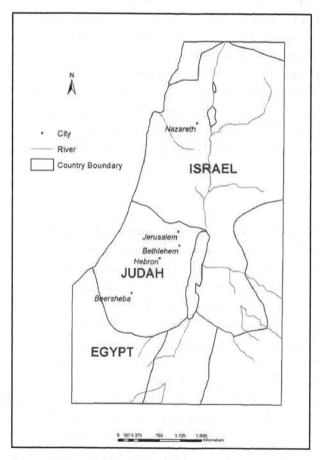

Figure 7.5　The Early Diffusion of Judaism.

(University of Akron produced)

CHRISTIANITY

The Christian religion in its various guises is the most widespread religion in the world, with more adherence than any other religion. It is one of three religions that grew out of the same geographic area and it shares many spiritual and fundamental teachings with Judaism, while Jesus is regarded as a prophet among Muslims.

The Christian Church has grown over the last 2,000 years to encompass one-third of the world's population, yet is the most divided of the major religions especially among the Protestant branch. The main divisions are the **Roman Catholic, Eastern Orthodox** and **Protestant** or Reformed Christian Churches. The latter includes many diverse groups from Anglicans (Episcopalians) to evangelists. Each Christian sect distinguishes itself from the next based on their theological beliefs and practices yet most will have common agreement in their desire for future redemption embodied in the sacrifice of Jesus Christ.

ORIGINS OF CHRISTIANITY

Christianity originated over 2,000 years ago essentially as a radical movement within Judaism. The movement grew around the followers of Jesus, a radical Jewish religious teacher. Jesus took great pains to point out the failings of the Jewish religion and especially its leaders. His followers, or disciples, regarded him as the messiah, the Christ, or God incarnate. The message he provided of the coming of the Kingdom of God on earth together with an ethos of faith, humanity and love for one's fellow human beings, became the basis of the Christian worldview. This viewpoint was later augmented and expanded by the disciples to incorporate beliefs about Jesus' own life; his birth, baptism, crucifixion and resurrection. The disciples also included stories about his public ministry and performance of several miracles in the area of Galilee into their teachings and ministries. The recognition of Jesus as the incarnation of God who died to atone for the sins of humanity is central to the Christian worldview. Along with these stories or testaments the Christians still adhered to the fundamental principles of Judaism; recognizing the same God and the words of the prophets as written in the Tanach or Hebrew bible. These scriptures form the Old Testament of the Christian bible while the writings of the apostles make up the New Testament.

THE DIFFUSION OF CHRISTIANITY

In the early days of the movement there was great unrest in the area of Palestine. The Jewish people were growing evermore restless with the oppression under Roman rule and the confusion among the various sects, Sadducees, Pharisee and Zealots. The new movement did not adhere to strict Jewish principles and professed non-violence which in a time of great upheaval attracted many moderates. The movement was especially attractive to the Hellenized Jews, those people of Asia Minor who had succumbed to all things Greek prior to the Roman occupation and still embraced these cultural differences. The word of the Christian movement soon spread across Asia Minor with missionaries who were seeking converts in many parts of the Empire. One of the leading exponents of the new religion was a man named Paul, who through his writings and preaching was able to articulate the Christian message through much of the eastern region of the Roman Empire. Following the initial upsurge in the hearth region there was a notable decline in the number of converts after the 1st century. However, the religion spread rapidly to other parts of the Mediterranean

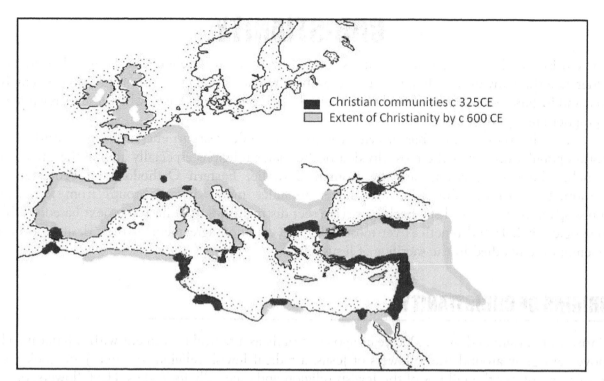

Figure 7.6   The Early Spread of Christianity. (© 2011, Leonard Peacefull)

during the later centuries. Despite initial fragmentations the movement became a Church with the acceptance of the belief by all Christians in the unity as one body of Christ.

In the 4th century further developments took place that would spread the faith throughout the Roman Empire and beyond. In 324 CE the Emperor Constantine and his family converted to the new religion and established Christianity as the official religion of the empire. This gave credence to the religion, a boost to its cultural acceptance and the consequent conversion of more people. Up to this time Christianity had spread throughout much of the eastern provinces of the Empire and could be found in pockets in the west. Now it was able to be freely spread to the most western extremities including Ireland. In the years that followed to cement this official recognition, several ecumenical councils set about defining the religion, accepting the Holy Trinity of Father, Son and Holy Spirit and establishing a hierarchy of bishops. However, political, economic and social problems in Rome were to act upon the religion and bring about its first split.

The Schism Between East and West

The Roman Empire had many differing culture groups but for much of its existence it had essentially been divided between the Latin and Greek traditions and languages. With the growing problems of the Empire by the end of the 4th century the split between east and west was final. The divergence had begun at the end of the 3rd century when Diocletian had sub-divided the Empire making his eastern capital Nicomedia (now Izmit in Turkey), leaving the western portion of the Empire to succumb to invading Germanic hordes. The separation of the Roman Empire into two factions had an impact within the Christian church. From the earliest time the bishop of Rome—

later known as the pope—had been *primus inter pares*, first among equals, but with the relocation of the capital to Byzantion (later Constantinople) there were now two patriarchs, the bishops of Constantinople and Rome, both of whom tried to impose ultimate jurisdiction over the church. With the western part becoming more isolated the bishop of Rome became more independent having no imperial control while in the east the bishop of Constantinople was part of the imperial machinery.

These two factions lead to distinct ideologies and important differences in beliefs and practices. The use of icons in the practice of worship became a major cause of disparity. The Eastern Church developed along diverse lines from the west; monasteries were dependent on the laity, celibacy was only required of the bishops, while the growing of beards was encouraged to develop an affinity with Jesus. In the west, monasteries were self supporting, celibacy was imposed on all ordained members of the church and beards were frowned upon. In theological terms the differences revolved around the interpretation of Nicene Creed and the liturgy, which provoked strong views on both sides. The differences finally came to a head in the 11th century when both branches excommunicated the other. Although the disputes seem petty they were enough for the two branches to once and for all go their separate ways. Two distinct churches emerged both calling themselves catholic or universal representing all Christians; the Roman Catholic Church recognizing the authority of the pope and the Orthodox Church or Eastern Orthodoxy recognizing the patriarch of Constantinople as its head. The latter also recognizing that there were region patriarchs and that services could be conducted in the vernacular language, whereas the Church of Rome chose Latin for its services. The Eastern Church progressed along independent national lines. The ancient patriarchs of Constantinople, Antioch, Jerusalem and Alexandria continued to be held in high esteem, but gradually national independent churches began to emerge with their own bishop as head. In the 16th century, the Bishop of Moscow was raised to the rank of patriarch and still today enjoys a great amount of influence in the Orthodox community of Russia and Eastern Europe.

THE REFORMATION AND PROTESTANTISM

Roman Catholicism spread though most of Western Europe from the fall of the Roman Empire until the early 16th century when several pockets of dissatisfaction with Rome's doctrine emerged in Northern Europe. These reformers had various grievances that they wished Rome to address but their concerns were not welcomed. This period of reform coincided with the age of enlightenment in Europe when scholars were questioning the authority of the Bible as the one true source of knowledge. The Reformation movement generally opposed the practice of confession, monasteries, celibacy amongst the clergy, the concept of purgatory and the veneration of saints and Mary (Stump, 2008).

From about the middle of the 16th century to the middle of the 17th century many branches of Protestantism developed. It is worth noting the cultural and geographic underpinning of the reformation. Geographically the movement was founded in northern Europe and focused on the Germanic languages of this region and were led by such men as Martin Luther (in Germany), John Calvin (in Switzerland), Jacobus Arminius (in the Netherlands) and John Knox (in Scotland). Culturally the people of Northern and Central Europe were more tolerant of new learning and were more apt to question. Politically this was a region where the ruling classes were rejecting the authority of Rome and the interference by the church in matters of state. This movement was led by Henry VIII, King of England, who renounced papal authority and was pronounced supreme head of the church and clergy in England. These heads of state were also able to replenish their dwindling royal treasuries with wealth confiscated from dissolved monasteries. This was also a time of serious cultural change, the rise of the merchant class, an age of new discoveries and the appeal of the reform movements in the various states especially where individualism took precedence over universal piety. In the 18th century, a group led by John Wesley was to break away from the Church of England and form their own Church that was to become known as Methodism. Further developments occurred in the 19th century with the birth of the Church of Latter Day Saints (Mormons) and in the beginning of the 20th century with the advent of Pentecostalism. Both of these have their origins in the United States.

CHRISTIANITY TODAY

Apart from the United States, where all groups coexist, Christianity has a distinct geographic pattern across the globe. The Orthodox Church predominates in Eastern Europe, Catholicism is the religion of Southern Europe, Sub-Saharan African and Central and South America, while Protestantism, or Reform Christianity, can be found in Northern Europe, North America, Southern and West Africa, Australia and New Zealand. The Christian religion in its various guises has a definite impact within the cultures that subscribe to the teachings of their particular Church. It influences the lives of many both directly and indirectly.

It is not straightforward to define Christianity in a short section as here, because there are so many different groups all promoting different ministries while having conflicting views and beliefs. Many individual groups assert that they alone are the true Church and all others are in error of their ways. Here are some examples: Conservative Reform Christians believe that people are born out of sin and unless they are "saved" will, after death, be eternally tortured in Hell. Salvation, they say, is a gift of God and can only be attained by repenting one's sins. Roman Catholics also believe that salvation comes from God. However, they believe salvation is channeled through the sacraments to repentant persons. They also ascribe to Purgatory, a type of temporary Hell, where one goes to after death; the final destination, be it Heaven or Hell, depends on how the person performed during his or her life. On the other hand, Liberal Christians interpret hell symbolically, not as an actual place. They reject the concept of God having created a place of eternal torment; that would be profoundly immoral. He would not sink to such unethical levels. Progressive Christians do not believe in an afterlife but in the notion that all that survives our death is the influences we have on others.

In North America, Christianity has a profound influence on the culture and politics. Conservative Christian groups are opposed to many social/cultural freedoms such as abortion, equal rights and same-sex marriages. In education they are opposed to sex education and natural selection. Whereas, more liberal Christian groups have no objection to any of these ideas or their cultural impact and feel people are free to practice their religion however they so desire.

As a world religion, Christianity is one of the youngest yet has become the most widespread. It began as an offshoot of Judaism and has diffused across the world in a short period of time. The rapid spread is due to cultural adaptations whereby different groups in varied geographic locations promote different ministries. Many groups have conflicting views and beliefs based on cultural differences that have led to sectarian conflicts, to be discussed in a later chapter. One thing does remain the same, all Christian churches share the belief in the Holy Bible as the word of God and the resurrection of Christ as the path to salvation.

ISLAM

Islam, derived from the Arabic word **salam,** or *peace*, is the youngest of the world's great Religions as well as the second largest in numbers (refer to Table 7.1). About 21 percent of people follow this rapidly expanding religion. It is estimated that if this faith continues to grow at its current rate, it will become the most largely practiced faith in the world by the middle of the 21st century, outnumbering those who practice Christianity.

THE ORIGINS OF THE RELIGION

It is believed by Muslims that the origins of Islam date back to the beginnings of the world. This religion acknowledges many prophets who came to Earth as messengers sent from the divine. However, true Islam originated around 610 AD in what is now Saudi Arabia. In 570 AD, Muhammad Ibn Abdallah was born in the city of Makkah (modern day Mecca). According to Islamic teachings, while Muhammad was meditating at Mt. Hera, he received revelations from God through the archangel Jibril (also known as Gabriel). These words would come to be written down as the Qur'an, the holy book of Islam and Muhammad would come to be known as the great prophet.

The people who inhabited the Arabian Peninsula at this time were nomadic herders dependent on cattle for their living. Most of these Arabian tribes worshipped many gods, goddesses and idols. It sometimes surprises people that out of this nomadic culture a monotheistic religion like Islam would arise. However, we must consider that at this time the Arabian Peninsula was on the margins of two great monotheistic Empires; those of the Byzantines and the Sassanids. The capital of the Byzantine Empire was at Constantinople, or modern day Istanbul, near the Mediterranean Sea, while the Sassanian Empire was centered on Ctesiphon, near modern day Baghdad and encompassed both Iraq and Iran as far as the Indus River in modern Pakistan. Each of these powerful empires controlled large tracts of land for cultivation, routes of heavy trade and were led by bureaucracies and kings with large armies. In both Empires the state and religion were closely connected. Christianity was the Byzantine Empire's official religion while Zoroastrianism was predominant among the Sassanian Empire. Also in the region were Hebrew or Yiddish peoples who followed the Abrahamic tradition of the Hebrew prophets of Judaism. These monotheistic religious groups were challenged by the rise in Islam when Muhammad spread his word throughout their empires.

After he received his messages, Muhammad began to spread the revelations that God had bestowed upon him with the other citizens of Makkah. His first teaching was that the Arab people, who worshiped many gods and goddesses at this time, were to worship only one true god whom he called Allah. At the beginning of his ministry, many opposed his teachings and those who converted to this new religion of Islam were often humiliated, expelled from the community, beaten, tortured and sometimes executed. However, people continued to accept his message and those who adopted Islam began to flock around him. Before long his message had spread from the town of Makkah through the region of Arabia, something that troubled the Arab elders who wanted to maintain their traditional way of life. As a result, a plot to assassinate Muhammad was put into place and after receiving a message from Allah ordering him to relocate, Muhammad and about 200 followers migrated in the year 622 to Madinah (Medina). This journey known in Islam as the Hijra (Emigration) represents a crucial initial event. It is the beginning of an independent Islamic Community and marks the start of the Islamic Calendar and what would eventually become the center of the Islamic world.

Once Muhammad had settled in Madinah, religious converts began to flock to the city from every part of Arabia. This migration of Muslims to the city began to strengthen Islam's hold on Madinah and this allowed for the creation of an Islamic society and government. With Muhammad as their leader the Islamic state of Madinah grew in strength and was able to withstand and defeat an invasion from tribes who adhered to the older Arab traditions. By the year 630 CE Muhammad and the state of Madinah had grown strong so much so that Muhammad was able to take his now formidable army and force the people of Makkah to surrender and accept Islam. Muhammad rode into Makkah victorious and established the Haram and Kaaba as the most sacred space in Islam. Following this event the Muslim faith spread throughout the Arabian peninsula as thousands of Arab tribesmen converted to Islam.

THE ISLAMIC FAITH

The Islamic faith comprises of the sacred Islamic text called the Qur'an, the six articles of faith and the five pillars of faith.

The Qur'an is the Holy book of Islam containing the word of God as passed down to Muhammad over a 22-year period. The word is derived from quraa, which means to read or recite. The Qur'an consists of 114 chapters, *suras*, made up of verses, *ayat*. Since the time of Muhammad the Qur'an has been read, studied, recited and chanted aloud. It is believed that this text, being derived straight from God, is infallible and should be read only in Arabic. To be able to memorize the Qur'an is the greatest act of devotions a Muslim can accomplish.

The six articles of faith are the main doctrine of Islamic belief and they include:

- Belief in one god called Allah;
- Belief in angels that are comprised of light and serve different purposes;
- Belief in the Torah, Psalms and the Gospel, but the Qur'an in the only true book of Allah;
- All Muslims to believe in every prophet from Adam to Muhammad, including Jesus;
- Each Muslim believes that every human will be resurrected at the end of the world;
- The final article of faith is the belief in the divine creed, which refers to predestination.

The five pillars are the true expression of the Islamic faith, by which Muslims honor and worship Allah and the word of the Qur'an.

- *Shahada* is the most important pillar of Islam. It is the declaration of the Muslim faith that "There is no god but Allah and Muhammad is the messenger of Allah." This is the first prayer learned by Muslims and it is the first thing said to newborns and the last thing said to a person upon their death.
- *Salah*—the second pillar of Islam is prayer. Muslims believe that Allah indicated to Muhammad, during his Night Journey, that Muslim prayers must be said five times daily. Prayers may be said alone, or with others in a mosque. Group prayer is more blessed than individual prayer. Prayer must always be said facing the city of Makkah. The prayers are spoken in Arabic and reflect the message of Allah relayed by Muhammad, mandated in the Qur'an.
- *Zakah* is a word that means purification but has evolved to mean tithing and almsgiving. Muslims who are financially capable of doing so give a certain amount of money each year. The money should be given to care for the poor, sick and elderly, as well as to build and maintain mosques and places of scholarly learning as indicated in the Qur'an.

- *Sawn*—fasting, from sunrise to sunset, during the month of Ramadan. Ramadan is a holy month because according to Islam, during this time Allah revealed the first revelations of the Qur'an' to Muhammad. During this time, Muslims fast during the day to remember the suffering of the poor and hungry and to learn self-discipline, selflessness and giving. Muslims increase prayers and try to read the entire Qur'an during the holy month of Ramadan.

- *The Hajj* is the final pillar of Islam. It is a pilgrimage to Makkah undertaken roughly 60 days after the end of Ramadan and should be made at least once in the life of every Muslim who is physically and financially capable. In the Hajj pilgrimage, Muslims walk seven times around Kaaba, the sacred House of Allah and the holiest building of Islam, according to Ganeri, where they are to touch or kiss the Black Stone, which is in the southeast corner of the building and pray at Maqam Ibrahim, the boulder Ibrahim stood on when he rebuilt the Kaaba, according to Martell. Then Muslims drink from the well of Zamzam and proceed to visit the hills of Al-Safa and Al-Marwa in remembrance of Hagar's search for water for her baby Ismail. Next they travel on foot or by camel to Mount Arafat, the location of Muhammad's sermon on his final pilgrimage to Makkah. Muslims then travel to the hilly area of Muzdalifa, where they may spend the night in prayer in the desert, proceeding to travel to Mina, where they throw stones at three carved pillars, representing a rejection of the Devil, as Muhammad had done, according to Martell. This completes the holy pilgrimage and fifth pillar of the Muslim faith.

These pillars are practices or duties, which originated in the last sermon of the prophet Muhammad and are emphasized in the Qur'an. Islam has no organized "church" or a hierarchy of power, as there are no priest-like leaders or sacraments in the religion, but rather, these five governing pillars. Worship is conducted at home and in mosques and as a form of "lay theocracy," each Muslim is his or her own governing priest, carrying out the obligations of the Muslim faith through its five pillars.

THE DIFFUSION OF ISLAM

The diffusion of Islam proceeded rather rapidly after the death of Muhammad in 632. The cities of Makkah and Madinah, situated 450 kilometers apart in modern day Saudi Arabia, are considered to be the cultural hearth of the Islamic world. From here Islam spread throughout Arabia. This was the beginning of the religion's diffusion throughout the world. It would follow two distinct paths, either converting non-believers who chose the faith, or forcing those who have been conquered militarily to accept Islam as their true faith.

Muhammad's successors, the Caliphs (enlightened rulers), passionately followed his paths of religion, politics and military conquest. Muslims saw themselves as agents of *jihad* (effort or exertion), fighting evil and carrying the rule of Islam to the ends of the Earth. They respected the rights of Christians and Jews to indulge in their own forms of worship and sought ways to adapt non-believers to Muslim ways. As Muslims continued to trade and travel they spread Islam by word of mouth. As Muslim influence spread through political alliances they established educational centers to the word of God to the newly converted.

The first region to be solidly Muslim was Arabia itself then, by conquest, the former Byzantine provinces of Syria and Palestine. From here the word was spread north to Armenia, east into Persia and southwest into Egypt. In 637 CE a Muslim victory at Kadisiya in Iraq marked the beginning of the end for the Sassanian Empire. Conquest and subjugation by the successors of Muhammad, the Caliphs increased their influence so much that within two decades of his death his followers ruled from Yemen in the south of Arabia to Armenia in the Caucuses, from Egypt in the West to Iran in the East.

The geographic spread of the Muslim faith brought technological advances and learning. By 751 CE the Muslims had conquered their way east and met the Chinese coming from the opposite direction. A Muslim victory at the Talas River in modern day Kyrgyzstan established an eastern frontier but also allowed them to gain technological knowledge from the Chinese. For example, they learned to make paper from plant fibers. Travelers then took the knowledge westward to cities such as Damascus and Baghdad and from there to Europe. Further incursions saw Muslim leaders move into parts of modern China along the Silk Road. There is a claim that a Mosque in Xi'an dates back to the 8th century though this is unsubstantiated. However, there is a strong Muslim presence in Xinjiang where the Uygurs, an ethnic Muslim group, have dominated since the days of the Mongol empire.

The Muslim incursion into North Africa started about the same time as their movements eastward. By the early 8th century Islam had spread mostly by overland trade routes across all of North Africa from Egypt and Sudan in the east to Morocco and Senegal in the west. Soon the wealth and cultural influence of the Caliphs spread along the coast from Alexandria in Egypt, Tunis, in Tunisia, to Fez and Marrakech in Morocco. Muslim traders and scholars came to influence the cultures of North and West Africa.

They did not stop there and by 711 CE the Moors were established in Spain. Moving northeast they sought to conqueror France but were defeated by the Franks at Tours in 733 CE. Several setbacks at the hands of the Emperor Charlemagne forced the Moors back into southern Spain where for the next seven centuries a glorious Moorish Muslim civilization and culture flourished. Cities such as Cordoba and Toledo became important for trade and scholarship. In Cordoba geographers created maps of the known world. While in Toledo the library boasted the second largest collection of scholarly works in all Islam. While in the same town, blacksmiths produced the finest steel that was fashioned into swords and weaponry.

Long before Muhammad there was a flourishing trade between Arabia and the Indian subcontinent. Initially these traders spread the word of Islam. Early in the years following Muhammad's death the Caliphs recruited Syrian warriors and sent them eastward to spread the word by conquest if necessary to the land of Afghanistan and Uzbekistan. There is a debate over when the Arab armies actually entered the Indian subcontinent. Some believe it to be as early as the 8th century while others mark 1001 as the time when Arab armies swept down the Khyber Pass and hit like a storm. Led by Mahmud of Ghazi, they raided just about every other year for 26 years. They returned home each time, leaving devastation in their wake. Then for nearly 150 years they more or less vanished. But the Muslims knew India was still there, waiting with all its riches.

When the Muslim armies returned in 1192 they encountered a purely Hindu culture, with Buddhism as a secondary faith. Both were gradually subjugated and converted to Islam. Under Muhammad of Ghor, the armies laid waste to the Buddhist temples of Bihar and by 1202 he had conquered the most powerful Hindu kingdoms along the Ganges. After Ghor's death in 1206, Turkish kings ruled the Muslim acquisition until 1397, when the Mongols invaded under Timur Lang (Tamerlane) and ravaged the entire region. However, this would all change in 1527, when the Mughal (Persian for Mongol) monarch Babur came into power. Babur was a complicated, enlightened ruler from Kabul; he conquered the Punjab and quickly asserted his own might over the Turkish kingdoms of northern India. This was the foundation of the Mughal dynasty, whose six emperors would comprise most influential of all the Muslim dynasties in India.

Babur's grandson Akbar extended the empire further, as far south as the Krishna River. Akbar tolerated local religions and married a Hindu princess, establishing a tradition of cultural acceptance that would contribute greatly to the success of the Mughal rule. Akbar's own grandson Shah Jahan, although spending much of his time subduing Hindu kingdoms to the south, did leave a great impact on the cultural landscape of northern India. The colossal monuments of the Taj Majal (his favorite wife's tomb), the Pearl Mosque, the Royal Mosque built in Islamic style and the Red Fort are all part of Jahan's legacy to the culture of India.

Islam traveled across the Atlantic with the trade in slaves during the 17th and 18th centuries. Although Muslim Arabs carried on the African end of the business they did not discriminate between the Animistic natives and those practicing Islam. Consequently, many African slaves transported to the Caribbean and North America took with them their Islamic faith. It is estimated that as many as one out of every five slaves brought to the United States was Muslim. As migration to the new republic continues into the 19th century many Muslims came of their own will. People from Syria, Lebanon, Jordan and in the 20th century, Palestine, north Africa and other parts of the Middle East ventured to the United States and Canada in the hope of economic prosperity and religious freedom. Many of the new immigrants were highly educated doctors, engineers and scientists.

THE SHI'IA–SUNNI SCHISM

With Islam foremost in the news in the 21st century, most people are aware of its main divisions; the Shiite and Sunni sects. Like most religions Islam has undergone a schism—a division or separation of the religion owing to differences of opinion—however, this split derives not from doctrinal differences but a political one.

After the death of the Prophet Muhammad, the question arose of who was to take over the leadership of the Muslim peoples. One group believed that the successor should be chosen from the group closest to the Prophet and so chose Abu Bakr, a member of Muhammad's tribe to be the first Caliph, who would provide secular and religious leadership. These people came to be known as Sunnis or traditionalists. A second minority group favored the Arab form of inheritance passing leadership down through the family of the Prophet and so chose Ali ibn Abi Talib, his cousin and husband of his daughter Fatima, as their first Caliph. These people were known as Shiites or partisans. The dispute that ensued was resolved when the Shiites accepted Abu Bakr as the Caliph and the Sunnis were to recognize Ali as the fourth Caliph. Even so the two groups remained divided over the concept of leadership.

The dispute came to a head when Husayn, the grandson of the Prophet, was elected to be the leader of the Shiites. Husayn's stronghold was Kabala (Iraq) where, in 680, Yazid, the leader of the Sunni faction, attacked him and inflicted a severe defeat. Yazid's army slaughtered the garrison including Husayn. This event cemented the divide between Sunni and Shiite Muslim. To this day the Kabala is one of the most sacred sites of Shi'ia Islam and Shiites regard the first month of the Islamic calendar as a period of mourning. A second sacred city of Shi'ia Islam is the city of Najaf (also in Iraq) where Ali is buried and a golden mosque was erected over the site.

Today Shiites represent about 15 percent of Muslims throughout the world. They are dominant in Iran, Iraq, Azerbaijan, Lebanon and Bahrain (see Table 7.3 on the next page). Sunnis accept that the first four Caliphs, including Ali, were the rightful followers of Muhammad. However, they don't grant the kind of divinely inspired status to their clerics that Shiites do with their imams. Shiites believe imams are descendants of the Prophet.

Sunni Islam dominates in most other Islamic countries. The center of influence has progressed from Medina (Madinah), to Damascus and later to Baghdad as Islamic culture flourished and grew. Along with this growth there arose some differences on the interpretation of Islamic law. The early scriptures relating to the word of Muhammad did not codify any specific laws to cover specific issues

Country	Sunnis	Shiites and Offshoots
Afghanistan	84%	15%
Bahrain	30%	70%
Egypt	90%	1%
Iran	10%	89%
Iraq	32–37%	60–65%
Kuwait	60%	25%
Lebanon	23%	38%
Pakistan	77%	20%
Saudi Arabia	90%	10%
Syria	74%	16% (Alawites)
Turkey	83–93%	7–17%
United Arab Emirates	81%	15%
Yemen	70%	30%

Table 7.3 The Distribution of Islam. (Congressional Research Service)

that arose with the spread of the Islamic faith. In the course of time legal scholars came to distinguish between **Shariah,** the divine law found in the Qur'an and *fiqh,* laws reached through human interpretation. Several schools of thought on the subject emerged; in Medina, were the traditionalists, while in Basra (Iraq) a more rationalist school came to prominence. As time passed four schools of thought have emerged to dominate with geographic consequences. The Hanafi School is found today in Jordan, Lebanon, Turkey and Afghanistan. The Maliki School in North Africa, the Hanbail in Saudi Arabia and the Gulf States and the Shafi School dominates in Egypt, East Africa, Yemen, Indonesia and South East Asia.

While Sunni doctrine is more rigidly aligned in accordance with those various schools, its hierarchical structure is looser and often falls under state, rather than clerical control. The opposite is true in Shiitism where the clerical hierarchy is more defined and, as in Iran, the ultimate authority is with the imam, not the state.

The differing interpretations of the Shariah have led to diverse geographical and cultural variations around the Muslim world. These diversities give rise to different customs, variations in contract and inheritance law, crimes and their punishment and dress requirements for women.

So Why Is There Tension between the Two Groups?

The answer is found in the spiritual and historical differences that may seem irrational to those who are uninvolved. Sunnis and Shiites believe in the Qur'an and Islamic law. But their interpretation and application varies. Shia Muslims believe that the Imam is sinless by nature and that his authority is infallible as it comes directly from God. Therefore, Shia Muslims often venerate the Imams as saints and perform pilgrimages to their tombs and shrines in the hopes of divine intercession. Whereas Sunni Muslims contend that there is no basis in Islam for a hereditary privileged class of spiritual leaders and certainly no basis for the veneration or intercession of saints. Sunni Muslims contend that leadership of the community is not a birthright, but a trust that is earned and which may be given or taken away by the people themselves.

There are inexplicable religious differences that are as impossible to solve, as is metaphysical analysis. People in the West, where peaceful societies depend on institutional mechanisms—hierarchical structures—that they have developed for channeling their differences of opinion away from violent conflict, find it difficult to comprehend these differences in Islam. The Muslim scholar Reza Aslan, in "No God But God," argues that those very institutional mechanisms are lacking in some Islamic societies such as Iraq, where the Sunni-Shiite divide is most pronounced. The battle going on within Islam today is defined by the struggle for control of those institutions especially in war-torn Iraq and Shiite dominated Iran.

Religious conflicts have continued throughout the centuries and it is not just within the Muslim faith. Christianity, with its various divisions, Catholic, Protestant and Orthodox, has experienced secular conflict within the realm of the faith. Factions within the religion have exploited these secular differences sometimes violently. In some instances the violence is in retribution for past acts. In Iraq it was the oppression of the majority Shiites by the minority Sunnis of the Ba'athist party. A similar conflict occurred in Christian Northern Ireland where a faction within the Catholic section of the population—a minority religious group attacked the oppressing Protestant majority. For more on this discussion, see the topics in the section on Cultural Conflict.

KEY TERMS

Supreme Being	Rituals	Sharia law
Ethos	Morals	Worldview
Polytheistic	Semitic Hearth	Monotheistic
Vedism	Vedas	Indic Hearth
Artha	Karma (Kama)	Dharma
Buddha	The Four Noble Truths	Moksha
Proto-Judaism	Judaism	Torah
Sabbath	Kabbalah	Schism
Roman Catholicism	Eastern Orthodox	Protestant
Salam	Qur'an	Sunni
Shi'a		

DISCUSSION TOPICS

1. To fully understand another's culture a person must try to understand their beliefs and religion.
2. How far is the worldview of the Eastern religions different from those of the Semitic Hearth region (Judaism, Christianity and Islam)?

CHAPTER 8

Cultural Conflicts

Cultural conflict can be defined as a state of disharmony, or clash, between incompatible cultures, or peoples with opposing beliefs, ideas or interests. Throughout the world there are many forms of conflict that can be interpreted as being differences in cultural understanding. These conflicts can arise for many different reasons but the underlying causes can usually be linked to: ethnicity, ideology, economics, political or environmental issues. In some cases there can be a combination of issues. Many cultural conflicts are based on events in history that ignite and underscore current actions. To determine the status of the current feud it is oftentimes necessary to look at the history as well as the geography—to examine not just the symptoms but the cause.

Though many of these disputes may be local and may be settled by peaceful negotiation, some have become a focus for regional conflicts while others have escalated onto the international stage. Ethnic tensions have been a major source of territorial disagreement throughout history. For example, in recent times the problems in the former Yugoslavia between ethnic Albanians and Serbs in Kosovo, the military altercation between the Russian minority in parts of Georgia, and tribal differences in many parts of Africa have led to civil war.

Many of these disputes can be classified as either ethnic or ideological. In this chapter we will determine the difference between these two forces and then examine some of the major conflicts of the present and recent past. For the situations outlined it is hoped that the reader will undertake further research into the causes and resolutions of the conflicts.

Ethnicity

Culture determines the practices and beliefs associated with an ethnic group and provides for its distinctive identity. The term **ethnicity** is not easy to come to grips with. People define themselves as an ethnically distinct group when they have in common a set of customs and characteristics, a language, a history, or a religion. They usually share a common homeland. People of a certain ethnicity also share common ancestry, however diluted that may be. For example, many Americans consider themselves Italian even if their immediate ancestors have been in the United States for several generations and one or two family members may not have any Italian ties, yet they continue with the ethic traditions, speak Italian and share the same religion. Ethnicity is distinct from race. People of a certain race share a single biological identity but that does mean they are of the same ethnic group. In the United States the census bureau makes certain racial distinctions, as in Asian and African American, while people of Hispanic or Latino ethnicity can classify themselves as black, white or other.

Ideology

Ideology as a cause of cultural conflict arises through differences in ideas, attitudes, values, and beliefs that are held by members of a particular cultural group. More specifically, political and religious ideologies tend to be hotbeds of conflict. **Political ideologies** such as Democracy and Communism provide an ample supply of differing opinions and beliefs with which to fuel conflict as do the religious ideologies of Christianity, Islam and Judaism.

Sectarianism

Sectarianism is a form of ideology through which conflict can arise. Sectarianism can be described as the discrimination, prejudice or hatred arising from attaching importance to the perceived differences between subdivisions within a group, such as between different denominations of a religion or the factions of a political movement. A sectarian conflict usually refers to violent conflict along religious and/or political lines. Examples of sectarian conflict include the conflicts between Catholics and Protestants in Northern Ireland; the philosophical, political and armed conflict between different schools of thought seen between the Shia and Sunni Muslims; and the philosophical and political conflicts arising from the recent elections held in Iran.

Fundamentalism

Religious fundamentalists, as an ideology, can be found in various traditions such as Christianity, Islam and Judaism. Each teaches that there was, in the past, a "perfect moment"—some divine intervention or the teaching of that "moment or truth" which they fervently attempt to recover and reestablish. Cultural conflict arises when this attempt involves reacting to that which is seen as a threat to realizing the ideal—even if the ideal never actually existed. Examples of this type of conflict can be seen in the cases of selected Islamic groups (e.g., Hizbollah, Hamas and Islamic Jihad) who believe that corrupt leaders in predominantly Muslim countries are thwarting the realization of their vision of an Islamic state. Islamic fundamentalists also tend to believe that the enveloping dominance of external powers, most notably the United States, is polluting the Islamic culture. Christian fundamentalists see the words of the Bible as their perfect truth and desire that everyone should share and adapt to this truth.

ETHNIC CONFLICTS

Ethnicity as a cause of cultural conflict frequently occurs due to the opposing viewpoints of peoples who share any or all of the characteristics of country of origin, religion, and culture yet distinguish themselves as ethnically different. This situation manifests itself in several distinct forms. Many of the displaced peoples of Europe, the Roma for instance, share a different ethnicity/culture than the people of the country they inhabit. Today many ethnic groups wish to assert their independence and establish politically autonomous or semi-autonomous units or states. This increased political consciousness could have two outcomes—either consolidate the unity of states with homogeneous populations or break apart those states with diverse ethnic populations. What is clear in today's world is that few states are homogeneous and many are deeply divided.

The evidence that ethnic conflict is a worldwide phenomenon is abundant. The recurrent hostilities in Northern Ireland, Nigeria, Iraq and Lebanon, in the second half of the last century there was the Somali invasion of Ethiopia and the Turkish invasion of Cyprus; the mass civilian killings in South and Southeast Asia. Then there are terrorist activities by various ethnic groups such as the

Sikhs in India, Basques in Spain, Kurds in Turkey, Palestinians in the Middle East. The expulsion of groups based on ethnicity is something we rarely hear about, for example, the expulsion of Chinese from Vietnam, of Arakanese Muslims from Burma, Asians from Uganda, or Beninese from the Ivory Coast and Gabon. Sometimes ethnic tension can lead to riots such as those in China, India, Sri Lanka, Malaysia, Zaire, Guyana and a score of other countries. The list goes on; those mentioned illustrate the violent evidence of ethnic hostility. There are many less dramatic manifestations.

In country after country, political parties and trade unions are organized ethnically. There are movements to expropriate ethnically differentiated traders and expel long-resident workers of foreign origin. Armed forces are frequently factionalized along ethnic lines. Separatist referenda in Quebec and the Swiss Jura, a painful division of Belgium into zones for Flemings and Walloons, the protests of Berbers in Algeria all serve to mark the potent political forces of ethnicity in the

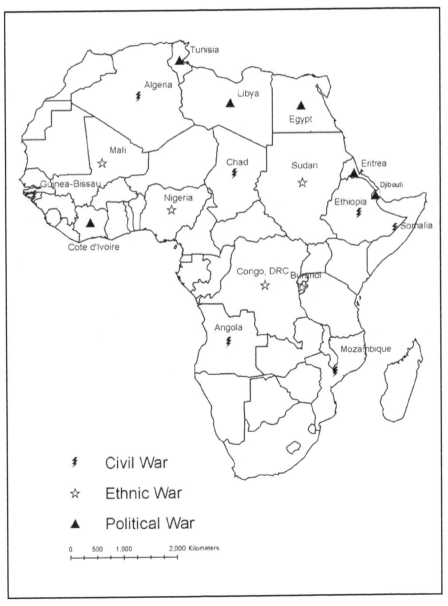

Figure 8.1   Africa, a Continent of Conflicts. (© 2011, Leonard Peacefull)

politics of both developing and industrialized states. In the next section we will deal briefly with some of the more recent occurrences and the main causes of the cultural conflicts.

CONFLICTS IN AFRICA

Africa is one of the most violent places where ethnic conflicts take on very dramatic and devastating elements. The map (Figure 8.1) shows the distribution of African conflicts. Much of the African turmoil of the post-colonial era is partly the result of the 19th century "carve-up" of the continent by the European powers Britain, France, Germany and Belgium. The arbitrary division of the continent, in the first instance, into governable units that later became politically independent countries by drawing lines through linguistic and cultural areas or by grouping different culture groups into the same country created the potential for conflict.

The Somalian Civil War

Currently there is no real national government in Somalia. While parts of the north are relatively peaceful, including much of the self-declared "Republic of Somaliland," interclan and **interfactional** fighting can flare up with little warning, and kidnapping, murder and other threats to foreigners can occur unpredictably in many regions.

Somalia is a product of the unification of two former colonies of Britain and Italy. Since independence in 1960 the country has been in state of continual turmoil. Fighting among rival faction leaders has resulted in the killing, dislocation and starvation of thousands of Somalis. Conflict between rival warlords and their factions continued throughout the 1990s and into the 21st century. No stable government has emerged to take control of the country. The UN assisted Somalia somewhat with food aid but did not send peacekeeping troops into the country. In the late 1990s, relative calm began to emerge and economic development accelerated somewhat. The country was by no means stable, but it was improving. A transitional government emerged in 2000 but soon lost power. Somaliland and Puntland, two regions in the north, broke away from the country and set up regional, semi-autonomous governments. They are not internationally recognized.

Since the attacks of September 11, 2001, Western governments have tried to bring stability to Somalia for various reasons. In January 2004, two dozen or so warlords reached a power-sharing agreement after talks in Kenya. This Transitional Federal Government (TFG) was the 14th attempt at a government since 1991. There has been continued heavy fighting with no real outcome. Meanwhile government and anti-government forces continue to fight each other. There are suggestions that Al Qaida is predominantly behind much of the unrest. Meanwhile it is reported that since 1991, 350,000–1,000,000 Somalis have died because of this conflict.

Because the country is so poor and without a strong central government, many young Somalis have taken to piracy, attacking shipping off the coast and far afield in the Arabian Sea and Red Sea. These hijacked ships bring in a large ransom from the ships owners, which support this very depressed economy.

Ethnic Conflicts in North Kivu

Since July 1994, an armed conflict between Zairian and Banyarwanda ethnic groups in the North-Kivu region of eastern Zaire has been ongoing. The conflict is being waged between Hunde and Hutu ethnic groups. Most of the clashes are concentrated in Masisi district, although sections of the district that have never before experienced tensions or violence have been engulfed in—or by—the conflict. Fighting is not only spreading to parts of the Walikale District of North-Kivu but is spilling over to South-Kivu, the neighboring region.

The root causes of the conflict stretch back to pre-independence, when Belgian colonizers initiated a well-organized migration scheme which brought thousands of Banyarwanda families, especially Hutu, from Rwanda to North-Kivu in order to develop the region. By independence, the Banyarwanda had become the majority ethnic group in their adopted country and Zairian ethnic groups, such as the Hunde, Nande, Nyanga and Tembo, were now a minority in their country of origin.

The Banyarwanda eventually gained economic and political power on both the local and national level, which became a threat to the traditional power structures in North-Kivu. The problem lies with the chiefs who, for the most part are Hunde and traditionally hold a great deal of political power as well as owning title to land in Masisi. These chiefs are also accused of extortion through irregular taxes and fines, demanding agricultural goods and livestock in exchange for the use of land plus arbitrary arrests and beatings for those who do not comply.

Several new factors, however, have complicated the nature of the conflict: the arrival of nearly a million refugees into the area in 1994, an increase in arms trafficking, the proliferation of youth militias and the presence of several hundred Zairian soldiers throughout the region. Today, the underlying motivation for the conflict is as much opportunism and a desire for economic gain as it is ethnic divisions. The violence and destruction had detrimental effects on the regional economy, on the population's access to health and education and development initiatives.

Darfur

Darfur, which means land of the Fur, is a desert region in western Sudan about the size of France. The conflict is the result of many years of tension over land and grazing rights. Although the antagonists are Arab Muslims, on one side and groups from black Africa on the other, the conflict has been seen by many observers to be ethnic and tribal, rather than one of opposing religion. However, a United Nations report confuses the problem by stating that the various tribes under attack, the Fur, Masalit and Zaghawa tribes do not appear to have a distinct ethnicity from their attackers. Among the causes of this conflict are a combination of decades of drought, desertification and over-population. While many blame the incursion of the Baggara nomads searching for water for their livestock, these people were forced to take their livestock farther south, to land mainly occupied by Black African farming communities.

The conflict began in February 2003. On one side of the armed conflicts were the Sudanese military and the Janjaweed, a Sudanese militia group recruited mostly from the Afro-Arab Abbala tribes from the northern region of Sudan. These people are mainly camel-herding nomads. The other side comprises a variety of rebel groups, notably the Sudan Liberation Movement/Army and the Justice and Equality Movement, recruited primarily from agrarian non-Arab Fur, Masalit and Zaghawa ethnic groups. The Sudanese government provided money and assistance to the Janjaweed while publicly denying its support.

There are many estimates of casualties. Reports of violent deaths compiled by the UN indicate between 6,000 and 7,000 fatalities from 2004 to 2007. Some non-governmental organizations use 200,000 to more than 500,000; the latter is a figure from the Coalition for International Justice. By 2006 as many as 2.5 million are thought to have been displaced, with some living in makeshift camps and others living in neighboring Chad. The eastern areas of Chad have a similar ethnic make-up to Darfur and the violence has spilled over the border area, with the neighbors accusing one another of supporting each other's rebel groups. The Darfuri refugees say the Janjaweed patrol outside their camps and men are killed and women raped if they venture too far in search of fire-wood or water.

The conflict is still ongoing although much of the fighting has died down. With intermittent hiatus for peace talks that tend to have no result, there are still the occasional skirmishes and causal-ities are kept on the rise. At present the conflicts and human strife in Dafur have moved from the world's headlines. Despite this, there are still millions of people without adequate food, water and sanitation.

FIGHTING FOR A HOMELAND

In several places throughout the world diverse culture groups are struggling for a place, area or coun-try to call their own—an autonomous territory which they can inhabit and govern while not hav-ing to answer to another central government. We are witnessing a period of political fragmentation where groups are asserting their right to **political self-determination.** In most instances this inde-pendence zeal is based on the group's cultural framework. In this section we examine the fight for independence of three of these groups; the Tamils in Sri Lanka, the Kurds in Turkey and parts of the Middle East and the Basques in Spain.

Sri Lankan Civil War

Sri Lanka has been mired in ethnic conflict since the country, formerly known as Ceylon, became independent from British rule in 1948. A 2001 government census states that Sri Lanka's main eth-nic populations are the Sinhalese (82 percent), Tamil (9.4 percent), and Sri Lanka Moor (7.9 per-cent). In the years following independence, the Sinhalese, who resented British favoritism toward Tamils during the colonial period, disenfranchised Tamil migrant plantation workers from India and made Sinhala the official language. In 1972, the Sinhalese changed the country's name from Ceylon and made Buddhism the nation's primary religion. As ethnic tension grew, in 1976 the Liberation Tigers of Tamil Eelam (LTTE) was formed under the leadership of Velupillai Prabhakaran. More commonly known as the Tamil Tigers, the LTTE campaigned for independence for the island's Tamil minority and the formation a Tamil homeland in northern and eastern Sri Lanka, where most of the island's Tamils reside. In 1983, the LTTE ambushed an army convoy, killing 13 soldiers and triggering riots in which 2,500 Tamils died. This was the beginning of a civil war that was to last for over 20 years.

In February 2002, the LTTE and the Sri Lankan government reached a cease-fire agreement, but both sides repeatedly violated the truce. After several on-off political solutions and cease fires, the assassination of Sri Lanka's foreign minister, Lakshman Kadirgamar in August 2005, reignited the conflict. For the next two years, both the government and rebels repeatedly violated the cease-fire agreement. In 2006, the government launched a military campaign to root out the rebels, and by July 2007, it had regained control of the country's east. From early 2008 the government pursued a fierce military offensive against the rebels, and in February 2009, claimed to have come close to defeating the separatist group.

In May 2009, the government declared the Tamil Tigers were defeated after army forces over-ran the last patch of rebel-held territory in the northeast. The military says rebel leader Velupillai Prabhakaran was killed in the fighting. In November new parliamentary elections were held. Opposition parties formed an alliance to fight the elections. The new alliance includes Muslim and

Tamil parties and is led by former Prime Minister Ranil Wickremesinghe. In April 2010 President Rajapaksa's ruling coalition won a landslide victory in parliamentary elections and in May the government said it planned to ease emergency laws in place for most of the past 27 years in response to its 2009 defeat of the Tamil Tiger rebels.

The civil war killed over 70,000 people and watchdog groups have accused both the Tamil Tigers and the Sri Lankan military of human rights violations, including abduction, extortion, and the use of child soldiers. The humanitarian problem of this conflict is the number of displaced persons that are left behind. Several NGOs and watchdog groups have accused both the Sri Lankan military and the LTTE of engaging in widespread human rights abuses, including abduction, conscription, and the use of child soldiers. In August 2007, Human Rights Watch released a report that catalogues alleged abuses on both sides of the conflict. Amnesty International has made similar accusations.

The Kurds

The Kurds are the largest cultural and ethnic group that does not have their own politically independent territory. Even though they have an ancient history the exact origin of the Kurds is not fully understood. With a total population of between 20 and 25 million, the Kurdish people are spread across six countries of Southeast Europe and the Middle East. It is a large territory principally in the mountainous regions of southeast Turkey, northeast Iraq, and northwest Iran; as well as Syria and parts of Russia. This is an area vernacularly known as Kurdistan, or land of the Kurds. Up until 1914, Kurdistan was divided between Iran, Russia, and the Ottoman Empire. After World War I much of Soviet Kurdistan was annexed to Turkey while a part of it was placed under Syrian rule and another under Iraqi when the Mosul region was annexed to Iraq. In Iran, the Kurds mainly resided in the northern provinces of Kurdistan, Kermanshah an, and south of the Western Azerbaijan.

Kurdish culture

The Kurds who occupy the same region as Turks, Iranians, and Iraqis share the same religion, Islam, but in all other ways they are ethnically and culturally different from their neighbors. It is thought that the Kurds share their origins with the people of Iran. The Kurdish language is a member of the Indo-Iranian group part of the Indo–European language family. However, because of their mountainous existence the language has become fragmented into many dialects that are often not always understood between various Kurdish clans. To confuse matters, written Kurdish may be in Latin, Cyrillic, and even Arabic scripts.

Turkish and Iraqi Kurds are predominantly Sunni Muslims of the Shafe'i, and some are followers of Yazidi and Ahl-e Haq sects, while Iranian Kurds are members of the Shi'ite sects, mostly Sufism.

Historically, the Kurds are a tribal culture group, with an economy based on nomadic herding. Tribal lands are the property of the chieftain who in many instances did not live on the land. Most thought of themselves not as Kurds but as members of one clan or another who regularly waged war on an opposing clan (Kinzer, 2008). The significant clans are Mokri in the north of Kurdistan and Bani-Ardalan more to the center around Sanandaj. The Jaaf are in southern Kurdiatan and the Kalhor toward the border with Kermanshahan. Today most Kurds live in small farming villages where they practice agriculture and animal husbandry. The settlement of Kurds in town is a recent phenomenon. In the Kurdish region of Turkey there were hardly any Kurdish towns. Towns like Diyarbakur, Mardin, and Siirt had populations of Ottoman officials, Arabic speaking traders, and

Turkish Armenians. While Kurdish chieftains had their mansions in town, the tribesman only came into town to trade their produce and buy necessities (Mango 2004).

The Struggle for an Independent State

The Kurdish people have fought through repression and have come close to having their own territory many times only to see these countries take away any chance of independence. Their struggles date back to the seventh century when they were first conquered by the Arabs and later by the Turkish Ottoman Empire. Their endeavors for autonomy and independence surged in the nineteenth century, when they were under the Ottoman Empire. It wasn't until after World War I that the Kurdish people had their first chance at gaining their own nation when the Treaty of Sevres proposed a division of the Ottoman Empire to incorporate a Kurdish homeland. Unfortunately, this treaty was quickly rejected and replaced by the Treaty of Lausanne which invalidated the idea of allowing the Kurdish people to have their own state. Since this setback the Kurds have been fighting for a place that they can call their own. This fight escalated toward the end of the last century and the early part of the present century. Many people, Kurd and non-Kurd, have lost their lives in an ethnic conflict that has no military answer.

This struggle was not confined to the Kurds of the Ottoman Empire either as Iranian Kurds also rebelled against the central government in 1880. Again in 1946, the People's Republic of Kurdistan, led by Qazi Mohammad, was established in Iranian Kurdistan, with Mahabad as its capital, however this short uprising was crushed by the Iranian army.

Up until the beginning of the last century most Kurds lived either within the Ottoman Empire or Persia. Since the end of the First World War and the breakup of the Ottoman Empire, Kurds have found themselves divided between Turkey, Iran, Iraq, and Syria; and with the collapse of the Soviet Union, Armenia, and Azerbaijan, have thus had to interact with a variety of ethnic groups. Even now many Kurds would like to live in their own autonomous region.

The main protagonists in the struggle for Kurdish independence is the Kurdistan Worker's Party, known as the PKK, Parti Karkerani Kurdistan. Their activities are focused on the Turkish authorities; their goal is for the government to recognize the political rights and cultural identity of the Kurdish ethnic group in Turkey and the creation of a Kurdish homeland. The PKK was formed in 1978, and has at various times been labeled a communist as well as a terrorist organization. The group has established camps in the mountains of Northern Iraq, close to the Turkish border, from where they launch their attacks on Turkish forces.

Kurds in Turkey

Throughout history the Kurds have occupied the mountains of eastern Anatolia. They are a tribal nomadic people who have only recently changed their cultural habits and settled into mainstream Turkish life. It is only in the present century that Kurds have been able to become assimilated in modern Turkish society. Since 2002 many reforms have been put in place; most significantly the constitution has been amended to allow Kurds to speak, broadcast, and receive education in their native language. Still, there are factions who are pushing for an independent Kurdish state in Turkey.

The main force behind the push for Kurdish independence is the Kurdistan Worker's Party, known as the PKK, Parti Karkerani Kurdistan. It is a Kurdish separatist organization that has been in a struggle with Turkish government for the creation of an independent Kurdish state for over thirty-five years. The PKK claims that Kurds have been long denied basic human rights; among these are linguistic, cultural, and political rights. Their goal is for the government to recognize the political rights and cultural identity for the Kurdish ethnic group in Turkey and the creation of a Kurdish homeland.

Contrary to popular belief the PKK was formed in Ankara, the capital, in November, 1978; not in southeastern Turkey where many Kurds are located. The PKK founder, Abdullah Ocalan, now in jail for subversion against the state, has along with his group been labeled communist as well as terrorist. The PKK have established a camp in the mountains in Northern Iraq close to the Turkish border; it is reported to have been equipped with a field hospital, electricity generators, and a high number of their lethal as well as non-lethal weaponry. It is from here that they launch their attacks on Turkish forces.

In more recent years there has been a move away from an armed struggle to a political one. The Democratic People's Party, which speaks for Kurdish nationalism, polled 6.3% of the vote in the 2002 general election. Ironically more significant are results from the 2007 parliamentary elections, when the Kurdish population gave more votes to the Government AKP and elected twenty members of parliament (Peacefull 2011). Unfortunately for Turkey, the Kurdish issue is holding them back from advancing on the world stage; because of the trouble with the Kurds the European Union has suspended Turkey's application for membership. In truth, there are not many options open to Turkey for resolving the Kurdish issue. It cannot be resolved by military means, nor is it feasible to set up a separate Kurdish state. Perhaps the only answer is complete integration into Turkish society.

The Kurds in Iraq

Unlike their ethnic cousins in other parts of the Middle East, the Kurds living in Iraq today enjoy a semiautonomous political existence, with a self-governing region in Northern Iraq. However, things have not always been pleasant for Iraqi Kurds. During the Iran-Iraq war in the early 1980s the Kurds supported the Iranian cause. Saddam Hussein, the Iraqi tyrant, retaliated against the Kurds by attacking them with chemical weapons and poisonous gas, destroying hundreds of villages and killing many thousands of Kurds. The attacks were so brutal they verged on near genocide of the Kurdish culture in Northern Iraq. Further troubles befell the Iraqi Kurds in the aftermath of U.S. Operation Desert Storm in which the Kurds sided with the U.S. in the hope it might gain them some form of independence. Saddam Hussein again fought back killing even more with poison gas attacks. A total in excess of some 5,000 people were killed in Halabja alone. Together around 180,000 Kurdish people died. A few years later they were to achieve their revenge on Saddam. During the U.S. invasion, called "Operation Iraqi Freedom," the Kurdish military forces in Iraq played a key role in the U.S. mission to overthrow Saddam's regime. After the fighting was over and a new form of government was established in Iraq, they were able to contribute to the new constitution that was established in 2005.

The Kurdistan region in Iraq is largely unified and shares a common government with the same two political parties that fought harshly years ago. The Kurds chose a democratic form of government and set up their own civil and democratic structures. They built up a judiciary system with police and military forces. Today Kurdistan is one of the safest places in Iraq. Within this region the gender conflict created by Islamic extremists is rare. In fact, about 45% of the students enrolled at The University of Salahadin are women. Women are seen more as an equal and important part of society. For Kurds, human rights are greatly important. Before 2003 education for any Kurds in Iraq was limited. Today the Kurdish people celebrate their freedom from tyrants, oppression, and cruelty.

Summary

A completely independent state of Kurdistan would mean Iran, Iraq, and Turkey would have to give up territory and with the territory, resources such as oil. Oil is a difficult resource for a country to part with no matter what the situation. If Kurds in Iraq can come to consensus and use their influence as a swing voter group they may be able to bargain for autonomy from a parliamentary coalition government with the Sunnis in Baghdad. Turkey may be willing to grant Kurds more autonomy if it believed doing so would help its chances of entering the European Union. Iran might see fit to publicly grant Kurds a measure of autonomy as an act of goodwill to impress the international community and to take the focus away from its nuclear program. The realization of an independent Kurdish state may never come to fruition, but given time, patience, and a good deal of negotiating, some form of ethnic autonomy can be achieved. All sides should act rationally and peacefully in order for anything to be accomplished.

A completely independent state of Kurdistan would mean Iran, Iraq, and Turkey would have to give up territory and with the territory, resources such as oil. The realization of an independent Kurdish state may never come to fruition, but given time, patience, and a good deal of negotiating, some form of ethnic autonomy might be achieved.

Figure 8.2 Kurdish Areas. (Courtesy of CIA and Perry-Castaneda Library Map Collection, University of Texas at Austin.)

THE BASQUES

The Basques are a unique culture group who occupy an area of northwest Spain and southwest France and are thought to be the oldest ethnic group in Europe. The Basques homeland, known as *Euskal-Herria* (Land of the Basques), straddles the Pyrenees Mountains that border France and Spain. The region is geographically varied ranging from the glaciated ridges of the Pyrenees to the coastal plain along the Bay of Biscay.

The Basques are thought to have inhabited this region for about 5,000 years. According to a Basque saying, "Before God was God and boulders were boulders, the Basques were already Basques." Over the centuries they resisted invasions by Romans, Barbarians and Moors. In the early 16th century much of the territory in Spain came under Catalan rule.

Basque Culture

The dominant feature of Basque culture is their language known as Euskera. It is unique in having no other relation and belongs to its own language family. During the 20th century along with regional dialects the language was suppressed, causing a decline in the number of Basques speakers. Since the restoration of the monarchy in Spain the Basque country has officially become bilingual with signs, particularly road sides, in both Basque and Spanish. Religiously the region is devoutly Catholic. One curiosity of the culture that attracts thousands of tourists is the annual running of the bulls in the city of Pamplona, where bulls run through the streets and young males dare to touch them without getting injured.

Basque Nationalism

Since 1954 the militant organization *Euskad Ta Azkatasurra* (ETA) (Basque Homeland and Freedom) has been dedicated to complete independence from Spain. In 1979 the Basque country was given local autonomy through a regional parliament. Regional autonomy appears to have satisfied the aspirations of most Basques. However, ETA—although representative of only a small minority—has continued to fight for full Basque independence with a series of bombings, mainly against police and army personnel. Civilians have also been killed and some indiscriminate civilian killings have produced demonstrations against terrorism. Basque terrorist activity continues to remain one of the major threats to the stability of the Spanish government.

Basques in France

In France there are no official statistics on the numbers of Basques, although it is estimated that about 40 percent of the population of the Department of Pyrénées-Atlantique are Basque. During the Franco era (1939–1975) the numbers were inflated as many Basques and other Spanish exiles fled to France. In addition, some French Basques no longer speak the language, which has no official recognition, although some educational courses are conducted in Basque and there is a small Basque press.

ETHNIC CONFLICTS IN CHINA

China, which is about the same size as the United States, has the largest population of any country in the world, over 1.34 billion. With such a large population there is a great variety in its constituent members. The government recognizes 56 main ethnic groups plus another group of over 300 small ethnic groups. The largest ethnic group is the Han Chinese, which constitute about 91.9 percent of the total population. The remaining 8.1 percent are Zhuang (16 million), Manchu (10 million), Hui (9 million), Miao (8 million), Uygur (7 million), Yi (7 million), Mongolian (5 million), Tibetan (5 million), Buyi (3 million), Korean (2 million), plus other minor ethnic groups.

In addition to the ethnic groups there are seven major Chinese dialects with many sub-dialects. Mandarin (or Putonghua), the predominant dialect, is spoken by over 70 percent of the population. It is taught in all schools and is the medium of government. About two-thirds of the Han ethnic group is native speakers of Mandarin; the rest, concentrated in southwest and southeast China, speak one of the six other major Chinese dialects. Non-Chinese languages spoken widely by ethnic minorities include Mongolian, Tibetan, Uygur and other Turkic languages (in Xinjiang), and Korean (in the northeast).

We previously defined ethnicity as relating to a distinct group of people who share a set of customs and characteristics, a language, a history or a religion. In China, the Han think of themselves as ethnically distinct yet amongst them there are groups that have little in common, neither spoken language, customs, characteristics, nor religion; while the members of the other 57 ethnic groups that make up the Chinese population do have languages, customs, characteristics, history and religions that separate them from one another and from the Han majority.

With so much diversity there is a great deal of cultural/ethnic conflict. So great is the impact of the many minorities that the Central government has established Autonomous Regions within the country, which enables the ethnic groups to have a certain amount of control over their daily life. Yet in regions such as Tibet and Xinjiang, home of the Uygur minority, there is a larger struggle for greater independence.

Xinjiang

Xinjiang lies in the west of China. It is twice the size of Texas but has a population of only 17 million, 60 percent of which is composed of at least five ethnic groups: the majority Uighurs, Kazakhs, Kirghiz, Tajiks, and Uzbeks. These groups are all Shiite Muslim and their languages are variations of Turkic. With the exception of the Uighurs, the minority nationalities extend across national boundaries to the north and west into Kazakhstan, Kyrgyzstan, Tajikistan, and Uzbekistan. These groups are all subject to the appeal of Muslim fundamentalism and an increasing desire to unite with fellow ethnic groups in the neighboring states. As a consequence, the last 25 years has seen organized and sometimes violent protests occurring with increasing frequency.

A movement to establish an independent East Turkestan Republic has attracted substantial support. In the 1980s, three groups were dedicated to this aim: the East Turkestan National Salvation Committee, the East Turkestan Popular Revolutionary Front, and the World Islamic Alliance. Over the following 30 years there were several uprisings, killings and bombings carried out by these various factions. Beijing has become concerned about international involvement in these incidents. The authorities cite evidence that Islamic militants in Saudi Arabia, Iran, Turkey, India, Pakistan, and Afghanistan have supplied weapons to Xinjiang separatists. None of the three central Asian republics support an independent East Turkestan, and along with the Russian Federation regard themselves to be threatened by Islamic fundamentalists.

One of the reasons for continued unrest is the central government's program of relocating ethnic Han from eastern China into the region. There are now large numbers of Han in Xinjiang who are mostly better educated and have higher skills. Some Uighurs complain that Han migration into the province has diluted their culture and marginalized them economically. The Han currently account for roughly 40 percent of Xinjiang's population, while about 45 percent are Uighurs. A majority of these in-migrants have secured better jobs and some have become entrepreneurs operating small businesses. This cultural division has caused unrest among many of the less fortunate young Uighur population. In several incidents in 2009 Han Chinese were killed or injured.

Tibet

Tibet is roughly one-third the size of the United States. It is geographically isolated and has historically been cut off from the outside world. Like Xinjiang, most of Tibet's population is non-Han (about 90 percent) while there are Tibetans living outside Tibet in portions of Sichuan, Qinghai, Gansu, and Yunnan Provinces. Tibet was an isolated country for centuries but in 1750 an agreement affirmed Chinese control over the territory. There was short lived independence in 1914 and this is the basis on which Tibet separatists claim their independence.

In 1950, after the establishment of communist rule, the Chinese People's Liberation Army invaded Tibet. For a period, things were relatively calm with only sporadic protests. In 1959 a series of protests expanded into armed rebellion, many Tibetans were killed and countless monasteries were destroyed. In the 1960s the Chinese central government vigorously undertook the "socialist transformation" of Tibet. Collaboration with Tibet's upper classes was replaced by a vigorous effort to mobilize support among Tibet's working classes. With the onset of the Cultural Revolution young men were discouraged from becoming monks, while most of the monasteries and temples were destroyed or defaced. Formal establishment of the Tibet Autonomous Region (TAR) in 1965 marked, paradoxically, the destruction of the last shred of Tibetan autonomy. China claimed it had accomplished the "liberation" of an impoverished feudal society. International human-rights advocates claimed that China was engaged in a systematic effort at suppressing Tibetan religion and the human rights of the Tibetan people.

Religion is a sensitive issue for the Chinese, and especially so in Tibet. The Dalai Lama is the spiritual leader of Tibetan Buddhists and has traditionally been responsible for the governing of Tibet. When China invaded in 1950 he left his official residence the Potala Palace in Lhasa, the capital, and fled to a refuge in Dharamsala, India. There is an ongoing dispute between Tibetan supporters of the Dalai Lama and the Chinese authorities that has often taken the form of protest demonstrations. The Dalai Lama would like to negotiate with Beijing to arrive at a formula for Tibet, not dissimilar from those in effect in Hong Kong and Macao, a "one country, two systems" approach. A degree of genuine autonomy that would guarantee religious freedom and local political control would satisfy many Tibetans. Other Tibetans want total independence and international recognition. Several questions come to mind in this regard. How would an independent Tibet sustain itself? Under what circumstances would the Chinese central government feel confident enough to grant genuine autonomy to a territory which it claims to have been always Chinese? How could Beijing give up a strategic barrier between China and potentially hostile states to the south? For the moment, the Dalai Lama, paradoxically, functions as a damper on ethnic separatism in Tibet.

THE HOMELESS ROMA PEOPLE

As a final examination of ethnic conflict we examine the fate of the Roma. The Roma people, sometimes referred to as gypsies, are a small ethnic group living in South Central Europe in countries such as Bulgaria, Czech Republic, Hungary, Romania, Serbia-Montenegro, Slovakia and Turkey. The population is estimated to be between 8.5 and 12 million as no real census has been undertaken. As each political unit in which they live has a different way of classifying its citizens, exact numbers are hard to come by.

The Roma people have not always resided in this hearth area of South Central Europe. Based on the similarity of their language and DNA testing it is thought that their original homeland was somewhere in Northern India. They first settled in the Middle East, calling themselves Dom, a word meaning "man." Records show they first appeared in Southern Europe in the 14th century and over the next 150 years spread throughout the rest of Europe. It is about this time that the name changed with the D becoming an R thus giving the word Rom.

Roma Culture

The Roma are split into several castes or clans, each with its own individual identity and ancestry. Because they are quite a diverse and widespread group there is not one definite explanation of the Roma culture. The Roma's have many customs and traditions that have changed over time as the people have adapted to the local prevailing cultural practices, yet they still tend to be conservative and follow the ways that have been practiced for centuries. For example, the women typically wear long skirts and dress more conservatively, only allowing a man's touch from their husbands. The language, Romani, is related to Sanskrit which is a member of the Indo-Europe language family. The religious commitment within the culture is very diverse. In the past they are reported to have been goddess worshippers, notably the goddess Kali. Although these beliefs have been abandoned there are still remnants of the old religion to be found within the culture, however, most subscribe to one of the predominant Christian religions.

Roma Persecution

Because of their cultural differences from the indigenous people, the Roma seem always to be met with hostility and repression and, as they continually do not live in one distinct location they have acquired the name Gypsies. Not belonging to one particular territory has caused them many problems across Europe. The Gypsies were immigrants, vagabonds who never stayed in one place for too long and appear to have been persecuted wherever they went. One of the most devastating periods in their history occurred during the Nazi Holocaust, in which an estimated 250,000 Roma victims were killed in death camps. In the present century Roma people have been subject to government persecution in many European countries, most notably Italy and France. It might be concluded that the problem has been exacerbated by the European Union's freedom of movement law (Schengen) that allows freedom of movement across borders; consequently a large proportion of Eastern Europe's Roma are moving west in search of employment. Italy, Spain and France have received the most. This influx of East Europeans has had serious social consequences including a rise in unemployment and crime in the areas the Roma inhabit. In 2009, the French authorities reportedly expelled 10,000 Roma back to Eastern Europe. The EU has demanded that states adopt strategies that increase the Roma people's access to education, employment, healthcare, housing and essential services in the hope of bridging gaps that currently exist between the Roma community and the rest of the population. Although these people are not waging an armed struggle for self-determination they are a group that is subject to persecution—a form of cultural conflict. They struggle for an identity and because of a basic cultural trait, one that has them continually on the move, they are constantly harassed and persecuted.

CONCLUSION

These are just some of the ethnic/cultural conflicts in the world today. Most, as we have seen, lead to armed aggression and/or terrorist activities that include bombing of civilian targets such the World Trade Center or Madrid Railway Station; or kidnapping and holding innocent people for ransom. Some, though limited in number, do reach more peaceful solutions. The fact remains that although ethnic conflicts usually develop into armed conflict, few are ever resolved in this way. It has become commonplace for many ethnic conflicts to be brought to a conclusion through the outside mediation of international bodies such as the United Nations or the African Union. However, one thing is certain that wherever two or more ethnic groups intersect, or where opposing ideologies meet, the conflict in one form or another is bound to occur.

CHAPTER 9

Gender Conflicts

Meera Chatterjee

Throughout the world there are struggles between the sexes. We refer to these as gender conflicts. In some areas these conflicts can be violent; in others they are more of a need for domination. In this section we will examine some of the struggles that reflect cultural altitudes on a worldwide basis.

Within underdeveloped nations around the world, a similar cultural theme occurs—the rising disconnect between genders and their role within their social order. In many developed nations, such as the United States, the gender gap in terms of rights and equality has been shrinking since women's rights campaigns swept the country politically and socially in the latter part of the 20th century. While in the west there is still a residual disjunction between genders, it cannot begin to compare to those in underdeveloped countries. These striking gender conflicts demonstrate a cultural diversity that separates them from developed nations. In order to clarify, we must first examine what these differences are. Next we will look at examples of gender conflicts concerning women, before finally observing how these examples define a culture.

Gender conflicts range from mere gender inequality to full on gendercide. In underdeveloped nations especially, these conflicts pose many social and political problems for its citizens. In many countries in the developing world the culture and the religion enhance male dominance. This dominance may lead to many differing and varying situations. For example, in many Muslim countries the women folk are obliged to cover themselves from head to foot when in public; in others their movements are restricted to the family dwelling. In Afghanistan, under the **Taliban,** young girls were denied an education. This still holds true in certain areas of the Muslim world. In contrast, in other parts of the world women hold dominant roles in the society. This can been seen in parts of West Africa and in the **matriarchal societies** of Easter Europe.

In many counties where women have less status, the sex trade industry is spiraling out of control. This industry gives scant regard to the safety and well-being of women who are forced into or are at risk of being forced into sex slavery, i.e., by having them participate in sexual acts against their will for profit of a third party. This industry is on the rise in many countries; none-more-so than in China.

WOMEN IN CHINA

While the **role of women** in China is undergoing great change as the country moves toward being an economic superpower, these changes bring with it certain consequences. Although China is a large country, about the same size as the United States, its population is five times as great. In addition, the geography is such that only the eastern third is suitable for habitation. The rest is either desert or mountainous with little opportunity to support large numbers of people. In the 1970s the central government realized that the population was growing to such an extent that before the end of the 20th century it would be difficult to sustain and that within a few years of the new millennium the population would be out of control. Consequently, a harsh new law was introduced to limit population; it is known as the **one-child policy.** In the initial years of implementation this policy, which applied only to the Han Chinese majority, restricted families to only one child. It was strictly enforced and in a society that is male dominated led to enormous social consequences.

Chinese culture is based on male dominance and **progeny.** The first-born male inherits and all males carry on the family name. Females are kept in very low regard. The male heir was needed to work in the fields or in the family business and then look after the parents in their old age. Thus, under the new law if the first born was female the family had to accept the fact or as more widely happened, the child mysteriously did not survive. In some areas women who had a girl child and became pregnant for a second time had two choices; either keep the newborn, if a male, hidden from the authorities, or have an abortion. This later practice was outlawed by the government in the 1980s but is still carried on under different circumstances today.

Besides outlawing forced abortions, the government also relaxed the limits on children in rural areas. If the first child was a girl then the family could have a second child. However, in urban areas the one-child limit still stands. This policy has many social ramifications. First there will be fewer members of the younger generation to take care of the older generation. Many urban children in future years will not know what a cousin is as their parents will have no siblings. Most significantly it has led to gender selection by the use of prenatal ultrasound and the abortion of undesired girl babies and may be a contributory factor in the large number of female suicides in the country, especially rural areas (see the section on Gendercide).

The maternal pressures are not the only factors bearing down on the majority of Chinese women. As most of China's population is considered rural, the role of women in Chinese culture is based on rural attitudes and male domination. In rural China women are so ill-respected they are seen as no better than a beast of burden. As a consequence of the custom that the man pays his bride's father a form of dowry for his new wife, he then feels he has bought her just like any livestock on the farm. There's a saying among men, "Marrying a women is like buying a horse: I have paid for you, therefore I can beat you and ride you whenever I like" Xie Lihua, editor *Rural Women*, magazine. (China From the Inside, 2006)

In addition, the new bride, upon entering her husband's family, has as a very low status; becoming no more than a drudge for her new mother-in-law. She may be beaten by her husband, by her father-in-law and under the old system, by her brother-in-law who may also, in her husband's absence, sexually abuse her. Needless to say, a woman so treated has few options. One is suicide, another is to run away. As a consequence, many rural young women move to the cities to find work and escape this harassed lifestyle. Unfortunately, they then become a target for a different kind of slavery.

Since the economic revival of the 1980s, many Chinese women now live the kinds of lives once unimaginable, enjoying good education, working for multinationals and owning their own homes. However, millions of their sisters, especially among the poor, have yet to see much change. There has also been a resurgence of many of the old attitudes and types of exploitation that the Communist Party sought to stamp out. The starkest example is the boom in the sex trade.

Although the government abolished prostitution in the 1950s, today massage parlors, hair salons and other venues offering sex for money have become ubiquitous; some estimates put the number of prostitutes in China at 4 million. Growth in this industry underlines how China's market economy, while creating new opportunities for some women, is contributing to the exploitation of others. There are several cases reported in the media that emphasize this exploitation. First there was the case of Deng Yujiao, reported in the *Guardian* newspaper in May and June 2009. Deng was arrested in Badong, Hubei, province for stabbing to death a local government official who she claimed had tried to rape her. When it became known that the 21-year-old waitress was to be charged with murder there was a huge public outcry in the media and online, coverage of her case in the state-controlled media was unusually sympathetic. In June 2009 the Badong County people's court exempted Deng from punishment for the crime of intentional injury because she was acting in self-defense, the official *People's Daily* newspaper said on its website. The report said the court ruled that Deng had limited criminal responsibility and also took into account that she turned herself in to police after using a fruit knife to stab her assailant. There is the case of five local officials in southwestern Guizhou who were jailed for forcing underage rural girls into the sex trade (Hewitt, 2009). Even though such exploitation is forbidden, other cases reflect a growing concern about the sex trade; while some of it is very true, some is over hyped. The authorities become overzealous in their drive to convict people on suspicion of involvement, as is the case of two schoolgirls in Kunming who were accused by police of working as prostitutes and then continued to press charges even after tests proved the girls were both still virgins.

These incidents have struck a powerful chord among ordinary citizens because of what they reveal about the status of women in China. While Beijing has officially promoted gender equality ever since Chairman Mao proclaimed that women "hold up half the sky," implementation of this ideal has proved patchy. The Chinese Communist Party made significant improvements in women's lives by granting them the right to divorce and to work on an equal footing with men and offering greater educational opportunities than those found in most other developing countries.

Even though China's constitution emphasizes that men and women are equal, it is very difficult to implement. Patriarchal attitudes still abound, as can be seen in restaurants where male customers will refer to female staff as "xiao mei"—meaning "little sister" or "little girl." The problem also lies in poor regulation and law enforcement. For example, it's very hard to prove that you've been discriminated against because of your gender.

In modern China there are signs of progress. Reports show that 50 percent of university students are women, which is a vast improvement in the last 30 years when there were only 23 percent in 1980. In other parts of society there are still huge gender gaps. For example, the average salary for white-collar men is 44,000 yuan ($6,441), compared with 28,700 yuan ($4,201) for women. Some women do succeed but there are limits to their success. Only 30 percent of senior managers in China's private enterprises are female.

China's women have always been under pressure: from men, from family, from work. Now more and more are under new pressure—from themselves—to take control of their lives; to get an education; to have a career; to marry for love. It's a slow, difficult process and it is changing China.

WOMEN IN AFGHANISTAN

The right of men to control women is not unique to China but may be found in many states in Central Asia, particularly in The Islamic Republic of Afghanistan where there is a fierce commitment to Islamic rule and the "enforcement" of gender specific laws. The articulation of these gender laws illustrates the domination of Islamic text in society whereby promoting any ideology that does not create a virtuous society is ultimately anti-Islamic. This has led to an extensive abuse of human rights, arbitrary and unlawful deprivation of the fundamental rights and duties of citizen and is advancing extensive ideologies against women and minorities.

Culturally among Afghan men there is a need to control all honorable resources: labor, land and women. Consequently, certain legislation aims to prohibit the equality among women in Afghanistan. For example, on August 17, 2009, a bill governing the familial life of Afghanistan's Shia minority, who make up 15 percent of the 30 million inhabitants, was secretly passed into law with the apparent approval of President Hamid Karzai. It permits Afghan husbands to withhold food from their wife if she refuses to submit to his sexual demands and requires women to have sexual relations with their husbands every four days. This law affectively condones and encourages "marital rape."

In contrast several decrees have attempted to fundamentally change Afghan attitudes. One was an attempt to alter the institution of marriage. The article forbids the exchange of women in marriage for cash or any other type of payment that is customarily due from the bridegroom on such occasions. These payments, from the groom to the bride—the *mahr*—are an essential part of the formal Islamic marriage contract. By implementing this law it basically gave women the liberty to choose who they wanted to marry. Unfortunately, the practice of finding a bride continues to force women into marriages with older men outside of their villages when they cannot find a close relative to enhance blood lines.

To suggest that the enforcement of gender constrains is purely political is misleading, as much of Afghan society is dominated by Pushtuns, who comprise 50 percent of the population and their cultural codes. Three core elements form the basis of this code, hospitality, refuge and revenge, which have been extended to include fundamental values are: equality, respect, pride, bravery, *purdah* (seclusion of women), pursuit of romantic encounters and the worship of Allah. Men have a higher status from their female counterpart; consequently, women are often treated exclusively as reproducers and pawns in economic and political exchanges. Women are subordinates dependent on their husbands and are expected by society to hand over their food, clothing and bodies in order to please their husbands who protect family honor. Sometimes the protection of the family honor can lead to some horrific scenarios with relation to women.

The Tragedy of Gender Conflicts in Afghanistan

Alyssa Bailey

The cultural attitudes regarding alleged violations of human rights fail to see the crisis in the development of modern Afghanistan. Discrimination laws in the constitution provide that any kind of discrimination and privilege between the citizens of Afghanistan are prohibited, yet local customs and traditions initiate prejudice. Violence against women is accepted if it helps promote a virtuous society in accordance with the principles of Islam (Warrick, 2005). If a woman's behavior is deemed immoral, to have impugned the honor of the entire family, then the men have the right and responsibility of defending this honor. This is then an appropriate condition to use violence against the woman in question. The man and his family group can, if their honor has been compromised or they are shamed, then look for a way to exact revenge for themselves and their family. Such honor killings are nothing more than murders, mostly based on suspicion, usually carried out by family members against a woman who is believed to have committed sexual indiscretion, or to have caused family gossip related to sexual behavior (Warrick, 2005). In an abundant amount of cases "honor killings" were premeditated and extremely violent: victims are not merely quietly done away with to restore family honor; instead they are killed with multiple stab wounds or gunshots, bludgeoning, or strangling, occasionally in public (Warrick, 2005).

The Afghanistan Independent Human Rights Commission reported on the case of a 29-year-old woman who had been sentenced to death by decree from a local religious scholar. On the grounds of Article 25 of the Constitution: "An accused is considered *innocent* until convicted by a final decision of an authorized court"; therefore, a local religious scholar upholds no authority to adjudicate legal cases. According to eyewitnesses, the 29-year-old, named only as Amina, was violently dragged out of her parent's house by her husband and local officials, before being publically stoned to death for committing adultery. The Afghan law protects individuals from this unconstitutional prosecution, arrest and execution, however, the law in action helped throw the first stone.

The socio-cultural tradition is reconciled with the legal system largely through the shifting implication of the law as it applies to cases where the victim is female. Despite humanitarian efforts, the failed attempt to stop the encouragement of gendered laws contributes to the ever-continuing repression of women in Afghanistan and the action of **self-immolation**—the voluntary sacrifice or denial of oneself, as for an ideal or another person.

Although the state has the duty to respect the liberty and dignity of human beings, as in Article 24 of the Afghanistan Constitution it does not execute this protection law, consequently women have no options other than self-sacrifice. Afghan women perform self-immolation: suicide by setting themselves on fire. In 2010, the Herat Burns & Plastic Surgery Hospital admitted 90 women who had self-immolated, 51 of them died, while in 2009 there were only 85 women who set themselves on fire but a greater number, 59, died. Most of those who self-immolate are between the ages of 15 and 25 (Quraishi, 2011). Esfandiari (2004) quotes Gurcharan Virdee, a worker with the Medica Mondiale who describes working with these women, "One of the most tragic cases involved a young pregnant woman who survived despite suffering severe burns over 60 percent of her body." Virdee goes on to describe another self-immolation case:

> "One of the women that I met, she was about 29. She already had four children, [and] she was seven months pregnant when she burned herself. She was experiencing problems with her husband and family; they wouldn't allow her to go and visit her own family. She set fire to herself. She then gave birth to a baby with no painkillers, nothing. The baby girl was taken by her aunt to look after her and the mother died three weeks after giving birth." *(Esfandiari, 2004)*

The criminal justice system is unwilling or unable to address issues of violence against women and minorities. Afghanistan's quality of protection against this demographic remains poor due to weak central institutions, a deadly insurgency and the country's ongoing recovery from two decades of war. The government continues to struggle in its endeavor to expand its authority over regions under the control of district commanders.

The complexities of abandoning human rights are delineated from the concept of de jure laws, how the law is written, versus de facto laws, what happens in practice. Essentially, the alteration of Afghanistan's Constitution from de jure, "liberty and dignity of human beings are inviolable," to the disenfranchisement of women and minorities who are subordinate to Islamic rule is not creating a

Republic of Afghanistan that the world envisioned. Afghanistan describes its laws as representative of a civil society free of oppression, atrocity, discrimination and violence, based on rule of law. From a citizen's perspective the struggle for power is between the Taliban and modern Afghanistan; where the innovative ideologies are being violated by Islamic extremists and their manipulation of the Islamic religion.

It can be disheartening to analyze this culture of corruption when the foundation of its society rests on the simple notion of Islam teachings: to live a morally perfect and peaceful life for God. There is no morality in slaughtering women and minorities for the sake of Allah or a peaceful life when one can be publically beaten and raped for "deviating" against an assumed norm, which is regarded as the standard of correctness in behavior by radical extremists. In order for the country to prosper, the conflict, corruption and negligence needs to be transmitted to the global community so that the world can understand.

WOMEN IN INDIA

Ever since humans have been on the planet, nature tends to maintain an equilibrium between the male and female populations. However, that biological law of nature has been challenged in India through their culture and more recently with modern medical technology. Indian society aligns along a very deep-rooted patriarchal pattern. Like any other culture of the world, males of Indian society made social and religious rules in their favor. And as India is an ancient civilization, these rules have been ingrained in the society for an extensive period of time. Paradoxically, women of India have also tended to believe in these rules for a very long time. Consequently, an exceedingly high degree of gender discrimination against females has been widespread throughout the country. The preference for "boys" over the "girls" is openly and commonly practiced in India. This results in a distorted sex ratio. Gender inequality is predominantly witnessed in education, employment and healthcare. Today though, education amongst women is bringing a wave of change as it makes them conscious about their place in society. However the change is based on the region, caste and economic condition of the women. The overall position of women is relatively better in urban areas as compared to villages. Currently, women's issues including discrimination are a combination of multiple dynamic factors which construct the diaspora in India.

India is one of the oldest civilizations of the world; the status of women over the course of history has changed many times. It has varied from a high status in the beginning of the ancient times to a low status in the medieval period, while modern India is experiencing an upsurge of equal rights resulting in a gradual escalation of the status of the women. During the ancient Vedic period, women held important positions in the society. The **Vedas** have allusions to women philosophers and intellectuals who were greatly appreciated for their spiritual insight. Later in history the caste system (social stratification) caused many internal conflicts within the society that affected women directly. Slowly over time the women often fell prey to internal dispute and the brutality of the alien aggressors. Consequently, the fortification of women became an imperative issue and men willingly annexed the responsibility. The apt female behavior was designed and laid down by Manu in 200 BC.

> "by a young girl, by a young woman, or even by an aged one, nothing must be done independently, even in her own house." "In childhood a female must be subject to her father, in youth to her husband, when her lord is dead to her sons; a woman must never be independent." (*Azad India Foundation, 2010*)

Consequently, the lives of women in India have been molded by the traditions that are thousands of years old. However, the situation has been changing in the past few decades. Like any other developing country, India also exhibits extremes in the placement of women in the society. In conjunction with traditional and conservative roles, women hold or have held high and responsible offices such as Prime Minister, Indira Ghandi, President of the country is currently Pratibha Patil and the Speaker of the Lower House, Lok Sabha, is Meira Kumar.

Skewed Sex Ratio

A very common Indian consecration for women is "May God bless you with a hundred sons." This is a direct expression of the Indian way of placing males over females. Indian "social security" works through the sons of the family; as the parents grow old they are supported by the sons. They are also treated as the prospective designer of family stature. Daughters on the other hand are to be married off to a different family. Consequently, they are treated as an economic liability. In a traditional Indian wedding, the bride's parents have to provide a handsome "dowry" to the groom and his family for setting up the new household. Additionally, the bride (daughter) is also given gold/silver jewelry to surge over financial difficulties in her future married life. Over time, these traditional elements have turned into an evil obsessive process and an extreme financial commitment for the bride's family. This practice is linked to the heinous crime of **dowry killings;** if the husband or his family is unsatisfied with the dowry or if the bride does not live up to expectations, then she is illuminated and the husband is free to remarry.

This social system forces people to perceive girls as a burden. Even today, the birth of boys is celebrated and the birth of girls is mourned upon. These social customs coupled with lack of education are directly responsible for a skewed sex ratio of the Indian population. The number of females in the population has been drastically declining since 1901 (Figure 9.1). Nevertheless, the last two decades show an optimistic development. Unfortunately in lower economic categories, the status of the women is ascertained predominantly by their childbearing function and their worth is frequently determined by the number of sons.

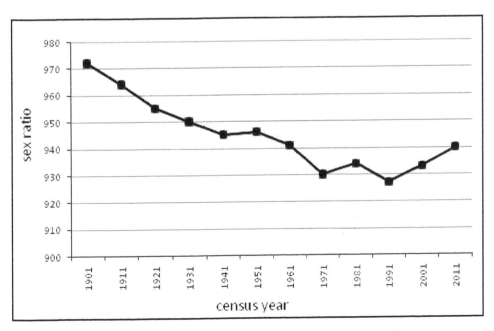

Figure 9.1   Sex Ratios in India Since 1901. (Census of India, 2011)

Female **infanticide, feticide** and sex selective abortion is shaping the abnormal sex ratio of India. Advancement of medical technology has allowed the prediction of the sex of the child before birth. However, this technology is utilized in negative ways by the people and medical practitioners. Both the parties are equally responsible for creating an abnormal sex ratio.

"When I was having my son back in 1991, I had to go through routine ultrasound. My doctor had let me know the sex of the baby. He was surprised to find that I was not over excited to have a son. In fact, he was concerned about my mental health" *(Author)*.

The number of female fetuses being aborted is rising in India despite the legal restrictions on using ultrasound to determine sex of the baby. Indian males ominously outnumber females and this has been increasing over time. According to the 2011 Indian census, the sex ratio stands at 914 women per 1,000 males. However, it fluctuates from state to state as the indicators responsible for creating such patterns vary (Figure 9.2). The sex ratio depends on the combination of social and economic factors. Two very prominent indicators are lack of education and poverty in rural areas. As the average level of education of females increases, the disparity in the sex ratio decreases. In a country like India, where skewed sex ratio is of great concern, Kerala, the southernmost state actually supports

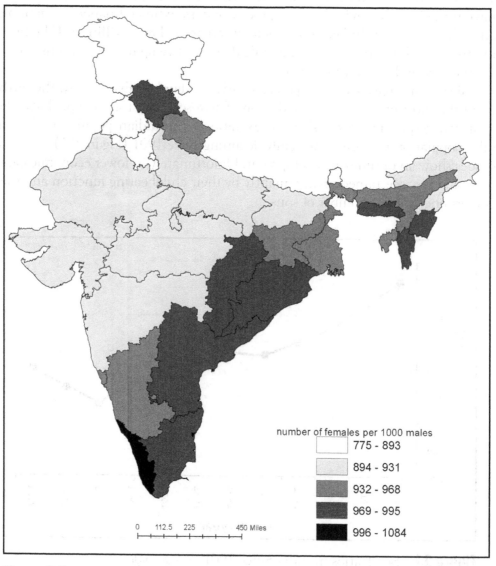

number of females per 1000 males

	775 - 893
	894 - 931
	932 - 968
	969 - 995
	996 - 1084

0   112.5   225   450 Miles

Figure 9.2   Ratio of Women to 100 Men. (Census of India, 2011)

more females than males (1,084 females per 1,000 males). This is attributable to the fact that Kerala has the highest female literacy rate according to the 2011 Census. The states with the least number of females are Punjab, Harayana, Sikkim and Jammu and Kashmir in the north. The overall education level of all the four states is low. Meanwhile Kashmir has also been embroiled in political turmoil for a prolonged period of time.

The government of India is adopting corrective measures to mend the damage done to the demography of the population. Following are some of the steps being taken to improve the sex ratio of the country.

- Education for girls in rural areas is the focal point of five year plans. An enormous amount of money has been allocated for primary education in India.
- Declaration of the sex of the baby by the medical practitioners has been made illegal.
- Sex selective abortion is unlawful.
- Awareness campaigns (Save the Girl Child) are organized all over the country.
- The dowry system, one of the root causes of such lop-sided sex ratios is being condemned socially and legally.

Consequently, some development and improvement can be seen. As an Indian woman it is a pleasure to observe the positive alteration of sex ratios in 2011 over 2001 (Figure 9.3). Most of the states

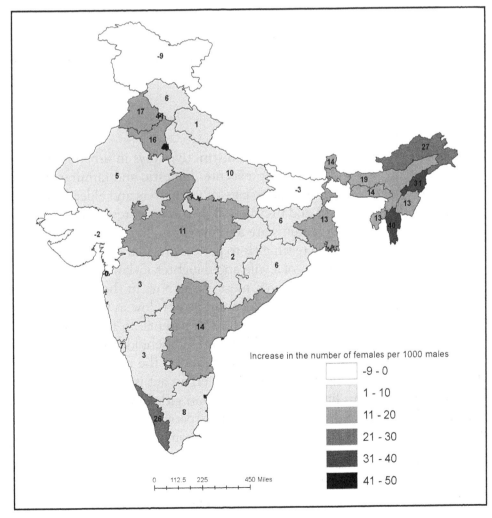

Figure 9.3   Change in Sex Ratio: 2001–2011. (Census of India, 2011)

have experienced an increase in number of females. Nonetheless, India is far away from the desired demographic balance between males and females. It is acknowledged that legislation alone is not adequate to deal with this problem that originates in social conduct and prejudices.

Food and Health

Women encounter discrimination from the beginning of childhood. Blatant gender disparity is evident in nutrition and education from infancy for a girl. **Malnutrition** is a very common cause of death among girls below the age of 5. Gender bias is one of the noteworthy responsible factors for creating such a pattern. Girls, in general, are breast-fed for a shorter period of time during infancy. Boys and males are fed better. Nutrition deficiency directly costs girls and women their health. Adult women in India consume fewer calories than required owing to the social prejudice. Women eat last and the least. This custom is practiced even when women are pregnant and lactating.

> "I have always seen my grandmother and rest of the women in the family eating the last and the least. Males ate first and bigger portion of food was fed to them. Leftover food from the day before was always eaten by the women of the house and fresh cooked was served to the males. This action was justified by the fact that they work hard and earn for the household." **(Author)**

It is reported that women avail less of the health care facilities than do men. This can be attributed to the fact that during childhood the illnesses of girls are all too easily neglected by their parents. Girls are socially trained and taught to tolerate aguish gracefully. Consequently, an adult woman is less likely to admit her sickness and neglect it beyond help.

Education

Gender discrimination in education is evident in distinctive ways in urban and rural India. In the Indian villages the young girls are expected to execute domestic and farming chores. Some of the important errands are cooking, cleaning, taking care of youngsters and elders, looking after the cattle, collecting potable drinking water and firewood. These sorts of responsibilities leave them with no time for education. Additionally, parents are also unwilling to spare them from daily chores for schooling. It is imperative to train and teach girls to perform regular chores from childhood as these are considered the prime responsibilities of a wife and daughter-in-law. Education has a limited role to play in the life of the village girls. However, most of the boys in the villages are sent to schools maybe up to the lower grades. As the overall education level is low in rural India, getting a suitable groom for an educated girl becomes very difficult for the parents. In the urban areas, however, the discrimination is visible in a different manner. There are households where boys are sent to private schools and girls to public. Boys only will be sent to school if the household has limited money to spare for education. The underlying justification for such an act is that boys need better education to provide for the family later in life. Nevertheless, urban girls have better opportunity to become literate and educated as compared to their rural counterpart. Following is an interesting narrative from an interview the author had with an urban Indian woman born into an educated family.

> "I am an US citizen, who was born in India a few decades after Indian Independence and was raised in a fairly large family of six sisters and a brother. My parents were more progressive than others and gave all of us children equal love and attention.

However, even in that situation, gender discriminations existed. My brother was given full freedom to play and participate in any sport of his choice. However, when I realized I had an interest in these sports and could do well in them, my father encouraged me to pick up activities or courses that were more feminine. Also, my sisters and I were strongly advised to pursue certain careers which are more represented as icons and carriers of tradition. We sisters were not allowed to travel alone or go far out of town for overnight trips without being chaperoned by a male, even if it was a younger brother.

In the workplace I had to fight for my rights to gain a higher position in the import/export business, which is traditionally a male dominated line of work. At home as well as at work, my opinions were second to those of any man, be it my husband, father, brother, a male co-worker or any male relatives. I had to work double hard to be taken seriously in a male dominated society.

Later on when I needed some legal consulting during a difficult season of my life, the legal system was not so supportive of a single woman fighting for rights under the constitution. I was not taken seriously and was advised by the authorities to reconsider my statements and unofficially asked to back off.

Like several other Asian countries, the preference of the boys over the girls is innate in India. Invalidating society's most fundamental traditions is an uphill task. It will take the joint efforts of government, media, youth and women themselves to be able to pull the society out of gender biasness.

GENDERCIDE: THE WAR ON BABY GIRLS

Technology, declining fertility and ancient prejudice are combining to unbalance societies:

XINRAN XUE, a Chinese writer, describes visiting a peasant family in the Yimeng area of Shandong province. The wife was giving birth. "We had scarcely sat down in the kitchen," she writes "when we heard a moan of pain from the bedroom next door . . . The cries from the inner room grew louder—and abruptly stopped. There was a low sob, and then a man's gruff voice said accusingly: 'Useless thing!'

"Suddenly, I thought I heard a slight movement in the slops pail behind me," Miss Xinran remembers. "To my absolute horror, I saw a tiny foot poking out of the pail. The midwife must have dropped that tiny baby alive into the slops pail! I nearly threw myself at it, but the two policemen [who had accompanied me] held my shoulders in a firm grip. 'Don't move, you can't save it, it's too late.

"'But that's . . . murder . . . and you're the police!' The little foot was still now. The policemen held on to me for a few more minutes. 'Doing a baby girl is not a big thing around here,' [an] older woman said comfortingly. 'That's a living child,' I said in a shaking voice, pointing at the slops pail. 'It's not a child,' she corrected me. 'It's a girl baby, and we can't keep it. Around these parts, you can't get by without a son. Girl babies don't count.'"

© *The Economist Newspaper Limited*, London, March 4, 2010.

In January 2010 the Chinese Academy of Social Sciences (CASS) said that within 10 years, one in five young men would be unable to find a bride because of the dearth of young women—a figure unprecedented in a country at peace. The number is based on the sexual discrepancy among people aged 19 and below. According to CASS, China in 2020 will have 30 to 40 million more men of this age than young women. For comparison, there are 23 million boys below the age of 20 in Germany, France and Britain combined and around 40 million American boys and young men. So within 10 years, China faces the prospect of having the equivalent of the whole young male population of America, or almost twice that of Europe's three largest countries, with little prospect of marriage, untethered to a home of their own and without the stake in society that marriage and children provide.

Gendercide—to borrow the title of a 1985 book by Mary Anne Warren—is often seen as an unintended consequence of China's one-child policy, or as a product of poverty or ignorance. But that cannot be the whole story. The surplus of bachelors—called in China *guanggun*, or "bare branches"—seems to have accelerated between 1990 and 2005, in ways not obviously linked to the one-child policy, which was introduced in 1979. And, as is becoming clear, the war against baby girls is not confined to China.

This is not wholly a Chinese problem. Parts of India have sex ratios similarly skewed while South Korea, Singapore and Taiwan also have peculiarly high numbers of male births. This gender discrepancy is not only confined to east Asia. Since the collapse of the Soviet Union former communist countries in the Caucasus and the western Balkans have also seen a rise in the sex ratio. Even subsets of America's population are following suit, though not the population as a whole.

The real cause is not any country's particular policy but the fateful collision between the overwhelming preference for a son, the use of rapidly spreading prenatal sex-determination technology and declining fertility. These are global trends. And the selective destruction of baby girls is global, too.

The Natural Order

In many societies boys are slightly more likely to die in infancy than girls. It is a natural fact that in order to compensate, more boys are born than girls. In all societies that record births, between 103 and 106 boys are normally born for every 100 girls. The ratio has been so stable over time that it appears to be the natural order of things. However, during the past 25 years the natural order has changed fundamentally. For example, in China the sex ratio for the generation born between 1985 and 1989 was 108, just outside the natural range. For the generation born in 2000–2004, it was 124. The ratio today is 123 boys per 100 girls. These rates are biologically impossible without human intervention.

The national averages hide astonishing figures at the provincial level. According to an analysis of Chinese household data carried out in late 2005 and reported in the *British Medical Journal*, only one region, Tibet, has a sex ratio within the bounds of nature. Fourteen provinces—mostly in the east and south—have sex ratios at birth of 120 and above and three have unprecedented levels of more than 130, "the gender imbalance has been growing wider year after year."

The BMJ study also casts light on one of the puzzles about China's sexual imbalance. How far has it been exaggerated by the presumed practice of not reporting the birth of baby daughters in the hope of getting another shot at bearing a son? Not much, the authors think. If this explanation were correct, you would expect to find sex ratios falling precipitously as girls who had been hidden at birth start entering the official registers on attending school or the doctor. In fact, there is no such fall. The sex ratio of 15-year-olds in 2005 was not far from the sex ratio at birth in 1990. The implication is that sex-selective abortion, not under-registration of girls, accounts for the excess of boys.

Not Only in China

Other countries have wildly skewed sex ratios without China's draconian population controls. Taiwan's sex ratio also rose from just above normal in 1980 to 110 in the early 1990s; it remains just below that level today. During the same period, South Korea's sex ratio rose from just above normal to 117 in 1990—then the highest in the world—before falling back to more natural levels. Both these countries were already rich, growing quickly and becoming more highly educated even while the balance between the sexes was swinging sharply toward males.

South Korea is experiencing some surprising consequences. The surplus of bachelors in a rich country has sucked in brides from abroad. In 2008, 11 percent of marriages were "mixed," mostly between a Korean man and a foreign woman. This is causing tensions in a hitherto homogenous society, which is often hostile to the children of mixed marriages. The trend is especially marked in rural areas, where the government thinks half the children of farm households will be mixed by 2020. The children are common enough to have produced a new word: "Kosians," or Korean-Asians.

China is nominally a communist country, but elsewhere it was communism's collapse that was associated with the growth of sexual disparities. After the Soviet Union imploded in 1991, there was an upsurge in the ratio of boys to girls in Armenia, Azerbaijan and Georgia. Their sex ratios rose from normal levels in 1991 to 115–120 by 2000. A rise also occurred in several Balkan states after the wars of Yugoslav succession. The ratio in Serbia and Macedonia is around 108. There are even signs of distorted sex ratios in America, among various groups of Asian-Americans. In 1975, the sex ratio for Chinese-, Japanese- and Filipino-Americans was between 100 and 106. In 2002, it was 107 to 109.

But the country with the most remarkable record is that other supergiant, India. India does not produce figures for sex ratios at birth, so its numbers are not strictly comparable with the others. But there is no doubt that the number of boys has been rising relative to girls and that, as in China, there are large regional disparities. The northwestern states of Punjab and Haryana have sex ratios as high as the provinces of China's east and south. Nationally, the ratio for children up to six years of age rose from a biologically unexceptionable 104 in 1981 to a biologically impossible 108 in 2001. In 1991, there was a single district with a sex ratio over 125; by 2001, there were 46.

Conventional wisdom about such disparities is that they are the result of "backward thinking" in old-fashioned societies. By implication, reforming the policy or modernizing the society (by, for example, enhancing the status of women) should bring the sex ratio back to normal. But this is not always true and, where it is, the road to normal sex ratios is winding and bumpy.

Not all traditional societies show a marked preference for sons over daughters. But in those that do, especially those in which the family line passes through the son and in which he is supposed to look after his parents in old age, a son is worth more than a daughter. A girl is deemed to have joined her husband's family on marriage and is lost to her parents. As a Hindu saying puts it, "Raising a daughter is like watering your neighbors' garden."

An Overwhelming Preference for Sons

In 1999 the government of India asked women what sex they wanted their next child to be. One-third of those without children said a son, two-thirds had no preference and only a residual said a daughter. Polls carried out in Pakistan and Yemen show similar results. Mothers in some developing countries say they want sons, not daughters, by margins of 10 to 1. In China midwives charge more for delivering a son than a daughter.

The unusual thing about son preference is that it rises sharply at second and later births. Among Indian women with two children (of either sex), 60 percent said they wanted a son next time, almost twice the preference for first-borns. This reflected the desire of those with two daughters for a son. The share rose to 75 percent for those with three children. The difference in parental attitudes between first-borns and subsequent children is large and significant.

Until the 1980s people in poor countries could do little about this preference: before birth, nature took its course. But in that decade, ultrasound scanning and other methods of detecting the sex of a child before birth began to make their appearance. These technologies changed everything. Doctors in India started advertising ultrasound scans with the slogan "Pay 5,000 rupees ($110) today and save 50,000 rupees tomorrow" (the saving was on the cost of a daughter's dowry). Parents who wanted a son, but balked at killing baby daughters, chose abortion in the millions.

The use of sex-selective abortion was banned in India in 1994 and in China in 1995. It is illegal in most countries (though Sweden legalized the practice in 2009). But since it is almost impossible to prove that an abortion has been carried out for reasons of sex selection, the practice remains widespread. An ultrasound scan costs about $12. In one hospital in Punjab, in northern India, the only girls born after a round of ultrasound scans had been mistakenly identified as boys, or else had a male twin.

The spread of fetal-imaging technology has not only skewed the sex ratio but also explains what would otherwise be something of a puzzle: sexual disparities tend to rise with income and education, which you would not expect if "backward thinking" was all that mattered. In India, some of the most prosperous states—Maharashtra, Punjab, Gujarat—have the worst sex ratios. In China, the higher a province's literacy rate, the more skewed its sex ratio. The ratio also rises with income per head.

In Punjab, Monica Das Gupta of the World Bank discovered that second and third daughters of well-educated mothers were more than twice as likely to die before their fifth birthday as their brothers, regardless of their birth order. The discrepancy was far lower in poorer households. Ms. Das Gupta argues that women do not necessarily use improvements in education and income to help daughters. Richer, well-educated families share their poorer neighbors' preference for sons and, because they tend to have smaller families, come under greater pressure to produce a son and heir if their first child is an unlooked-for daughter.

So modernization and rising incomes make it easier and more desirable to select the sex of your children. And on top of that, smaller families combine with greater wealth to reinforce the imperative to produce a son. When families are large, at least one male child will doubtless come along to maintain the family line. But if you have only one or two children, the birth of a daughter may be at a son's expense. So, with rising incomes and falling fertility, more and more people live in the smaller, richer families that are under the most pressure to produce a son.

In China, the one-child policy increases that pressure further. Unexpectedly, though, it is the relaxation of the policy, rather than the policy pure and simple, which explains the unnatural upsurge in the number of boys.

In urban China couples are only allowed to have only one child—the policy in its pure form. However, in the countryside, where 55 percent of China's population lives, everyone is permitted a second child either if the first is a girl or if the parents suffer "hardship," a criterion determined by local officials. All the ethnic minorities are excluded from the policy and are permitted more than one child regardless of the sex.

Not surprisingly, the provinces of the ethnic minorities are the only ones with close to normal sex ratios. These areas are inhabited by ethnic groups whose culture does not favor abortion and whose family systems do not disparage the value of daughters so much.

Guangdong, China's most populous province, has a sex ratio of 120, which is very high. But, if you take first births alone, the ratio is "only" 108. That is outside the bounds of normality, but not by much. If you take just second children, however, which are permitted in the province, the ratio leaps to 146 boys for every 100 girls. And for the relatively few births where parents are permitted a third child, the sex ratio is 167. Even this startling ratio is not the outer limit. In Anhui province, among third children, there are 227 boys for every 100 girls, while in Beijing municipality (which also permits exceptions in rural areas), the sex ratio reaches a hard-to-credit 275. There are almost three baby boys for each baby girl.

Things are not much different in India. First-born daughters were treated the same as their brothers; younger sisters were more likely to die in infancy. The rule seems to be that parents will joyfully embrace a daughter as their first child. But they will go to extraordinary lengths to ensure subsequent children are sons.

The Hazards of Sex Imbalance

Throughout human history, young men have been responsible for the vast preponderance of crime and violence—especially single men in countries where status and social acceptance depend on being married and having children, as it does in China and India. A rising population of frustrated single men spells trouble.

The crime rate has almost doubled in China during the past 20 years of rising sex ratios, with stories abounding of bride abduction, the trafficking of women, rape and prostitution. Studies show that higher sex ratios account for the rise in crime. In India, too, there is a correlation between provincial crime rates and sex ratios.

Violence is not the only consequence. In parts of India, where people pay a bride price (i.e., the groom's family gives money to the bride's), that price has risen. During the 1990s in China, there was a rise in inter-urban migration where couples from one-child areas who live in a legal limbo, move restlessly from city to city in order to shield their two or three children from the authorities.

Some of the consequences of the skewed sex ratio have been unexpected. It has probably increased China's savings rate. This is because parents with a single son save to increase his chances of attracting a wife in China's ultra-competitive marriage market. A study compared savings rates for households with sons versus those with daughters. It found not only that households with sons save more than households with daughters in all regions, but households with sons tend to raise their savings rate if they also happen to live in a region with a more skewed sex ratio. About half the increase in China's savings in the past 25 years can be attributed to the rise in the sex ratio, this would suggest that economic-policy changes to boost consumption will be less effective than the government hopes.

Over the next generation, many of the problems associated with sex selection will get worse. The social consequences will become more evident because the boys born in large numbers over the past decade will reach maturity then. Meanwhile, the practice of sex selection itself may spread because fertility rates are continuing to fall and ultrasound scanners reach throughout the developing world.

A Process of Reversal

At least one country, South Korea, has reversed its cultural preference for sons and cut the distorted sex ratio. South Korea was the first country to report exceptionally high sex ratios. Between 1985 and 2003, the proportion of South Korean women who felt "they must have a son" fell by almost two-thirds, from 48 percent to 17 percent. By the mid-1990s, the sex ratio began to fall and is now 110 to 100.

It takes a long time for cultural attitudes to change. Moving into a modernized world not only makes it easier for parents to control the sex of their children, it also changes people's values and undermines those norms which set a higher store on sons. At some point, one trend becomes more important than the other.

It is just possible that China and India may be reaching that point now. The Indian census of 2000 showed the sex ratio stable at around 120. At the very least, it seems to have stopped rising as shown in the provinces that had the highest sex ratios they have seen their ratios begin to decline. In India, the cultural preference for sons has been falling too and that the sex ratio, as in much of China, is rising more slowly. In villages in Haryana, grandmothers sit veiled and silent while men are present. But their daughters sit and chat uncovered because, they say, they have seen unveiled women at work or on television so much that at last it seems normal to them.

Though the two giants are much poorer than South Korea, their governments are doing more than ever to persuade people to treat girls equally. The unintended consequences of sex selection have been vast. They may get worse. At long last there seems to be a resolution to the phenomenon of Asia's "missing girls."

This section is based on an article in *The Economist*, March 6, 2010.

KEY TERMS

Matriarchal Societies	Role of women	One Child Policy
Progeny	Gendercide	Self-Immolation
Caste	Dowry Killings	Infanticide
Feticide	Malnutrition	Taliban

ACTIVITY

View the PBS video *China From the Inside* episode 2 at the following web site.

http://topdocumentaryfilms.com/china-from-the-inside/

After watching the video answer the following questions.

1. What pressures do Chinese women have that are not faced by Chinese men?
2. Do you see any ways that women wield power and in what ways are they rendered powerless?
3. Is there a connection between education and power?
4. Why do many parents in China favor having boys?

DISCUSSION TOPICS

1. Discuss the roles of women that are expressed in the video.
2. How are those roles being affected by the economic changes occurring in China?
3. Which of the beliefs match or challenge your own beliefs about gender roles?
4. Which views are most likely to provide women with basic human rights, and the opportunity to prosper?
5. Discuss the impact of the "one-child policy"? What alternative/s would you suggest to this policy to control the growth of population?

RESEARCH TOPIC

There are several websites that deal with women and gender conflict in Afghanistan, India and the Middle East beside those in China. Using the text and these websites, plus any other sources of information write an essay on *Gender Conflicts in the Developing World*.

CHAPTER 10

North American Cultural Diversity

North America's culture is part of a greater European Culture realm derived from the cultures of the British Isles and northern Europe. History tells us that whenever an empty territory undergoes settlement or an earlier population is dislodged, the specific characteristics of the initial groups are crucial for the later social and cultural geography of the area, no matter how tiny the initial band of settlers may be. Put another way, in terms of lasting impact, the characteristics of a few initial colonizers can mean much more than the contributions of tens of thousands of subsequent immigrants generations later. A case in point is seen in the northeastern United States where a small band of pioneers has greatly influenced the **Yankee culture.**

THE EARLY AMERICANS

Who were the early Americans, those who gave us the varied culture that we have today? In "*A Cultural Geography of the United States*" Wilbur Zelinsky (1973) recognizes five processes that combined to transform elements of the Older European culture into the current American one. We will try to catalogue the development of American culture using these five elements.

The Diffusion of Immigrants with Certain Cultural Traits

Early European settlement of the continent tended to be confined to specific areas. The Spanish had insinuated themselves into the south from their Middle American colonies; they settled in the Rio Grande area and northern Florida. The latter area grew around St. Augustine, a fortified town built to protect the treasure fleet as it sailed back to Spain. The Dutch had a small colony in the Hudson valley that was based on the fur trade. The French were the most widespread although they had colonies on the coast and in the St. Lawrence valley the main thrust of their colonizing of the interior of North America is symbolized by the **coureurs de bois,** French fur traders who ventured into interior and set up trading posts that later became small towns. Lastly came the British; the English, Welsh and Scots who settled the east coast in three phases.

First the **tidewater colonies** of the Carolinas, Virginia and Georgia were settled by wealthier land owners and minor members of the aristocracy who set up large self-sufficient estates or plantations. Later the Puritans, lower-middle class folk with strict religious beliefs, settled in villages and small towns in New England. The final group, a mixture of Quakers, Scots Protestants and Germans settled in the Middle colonies. Each one of these varied culture groups brought their own ingredients to what Zelinsky refers to as the lumpy stew of American culture.

Long-Distance Transfer of Peoples and their Cultural Baggage

The movement of large groups of peoples from Europe to North America resulted in the formation of a variety of different culture groups in the New World—New France, New Spain and even the New England. Although these colonies were all governed from their respective homeland they did however, initiate a different set of cultural patterns. The relocation of people across a great distance inevitably generates cultural change, however zealously the immigrants may try to prevent that change. Each of the new colonies had their own cultural ideals, baggage for want of a better term, that eventually became subsumed into the cultural patterns of modern America.

The Spanish settlements, in what is today northern New Mexico, along the upper reaches of the Rio Grande, was an extension of the Spanish colonies of Middle America known as New Spain. The area was initially colonized as a private venture by Juan de Onate and 210 other Spanish colonists, who brought with them not only their religion but also their customs, language and architecture. These new venturers established a ranching economy and went on to acculturate the local population. At the time the area had a considerable population of Pueblo natives, living in adobe-and-stone structures and practicing irrigation, who were prospects for the "missionary work" of the catholic priest who accompanied the new colonists.

The French were the first Europeans to establish permanent settlements in northeastern North America. Their initial settlements were in the maritime provinces of Canada but their more important settlements were in the St. Lawrence River valley from where the French spread their influence far inland. The vanguard of interior development was the *coureurs de bois*, who made their living in the fur trade, leading many water-borne expeditions far inland to explore and obtain their wares. During their voyages these men voluntarily adopted an Indian way of life. They relied upon native technologies, native languages, and the services (sexual as well as economic) of native women that eventually led to intermarriage and mixing of the cultures. It is the settlements pattern of this small but influential group of explorers, traders, missionaries and settlers that is reflected in such place names as Detroit, Mobile, New Orleans and St. Louis.

In the St. Lawrence Valley most settled on farms under the feudal seigneurial system, whereby large tracts of lands (seigneuries) were granted to favored individuals who were then charged with bringing in tenant farmers. These water-oriented colonists divided their seigneuries into long lots, properties with narrow riverfronts that extended perhaps 10 times as far inland away from these streams. The system can still be seen in many parts of the former French colonies in the St. Lawrence and parts of Louisiana. Although initially few in numbers (less than 10,000) with a high birth their numbers grew to such an extent they were, until 1834, the largest ethnic group in Canada. French-speakers now constitute about one-quarter of the Canadian population clinging tenaciously to their language and Catholic religion; they continue to exert a major influence on Canadian affairs.

The third group to settle in North America was the Dutch who established a small colony in the Hudson Valley, New York. The Dutch established a fur trading post at Fort Nassau, near present-day Albany, New York. From here they traded with the native Iroquois, who had direct and easy access to the west via the *Mohawk Gap* through the Appalachian Mountains. Although the Netherlands only controlled the Hudson River Valley less than 60 years, in that short time Dutch entrepreneurs established a series of trading posts, towns, and forts up and down the Hudson River that laid the groundwork for towns that still exist today. Fort Orange, the northernmost of the Dutch outposts, is known today as Albany; New York City's original name was New Amsterdam.

The highly varied colonies that were established by English colonists along the eastern seaboard proved very successful in attracting European settlers. Their physical environments ranged from the semitropical coastal lowlands of Georgia and South Carolina to the midlatitude shores of New England. These colonial settlements can be classified into three broad groups, based on their cultural and economic similarities. We can identify these cultural—economic regions as hearth areas of American culture from where their cultural influence spread to the rest of the country. These three hearths—the Tidewater colonies of the south, New England in the Northeast and the Midlands, around Pennsylvania—all developed different economies and cultures.

The tidewater colonies were the first of the English settlements, chosen because their similar latitude to the Mediterranean. It was assumed that the area could produce such Mediterranean products as grapes, citrus fruits, and silk, which were in demand in England. Experience soon proved otherwise. The first settlement, Jamestown, was poorly located in an unhealthy swampy area which led to a high mortality rate. With survival threatened, the venture was saved by moving to a more healthy location and by the discovery that tobacco, the "stinking weed," could be grown successfully, proving highly profitable, ensuring the colonies survival and growth. However, tobacco production required much labor, the bulk of which was supplied by indentured servants, poor young Englishmen who, in return for their passage to the New World, worked for their sponsor for four to seven years, after which they would receive help in becoming established on their own. This system had its problems, producing a great sex imbalance and social cleavages. Indentured servants were also deemed lazy by the plantation owners, which led the owners to look for other laborers and to the introduction of a new culture group to the Americas—the African slaves.

Tobacco and later cotton were to become the economic staples of this region. Both were grown on plantations and needed a large labor force. This was provided by slave labor. In the last 30 years of the 17th century more than 12,000 slaves were brought to the tidewater region. The plantation lifestyle produced a relatively self-sufficient culture producing most of the food, fuel, and other items. Skilled workers such as blacksmiths and coopers lived either on the plantation or close to them. Consequently there was little need for the services normally provided by towns. An overwhelmingly rural, agrarian, dispersed, and decentralized culture developed that can be seen as the foundation of the culture of the southern states of the United States.

The **New England hearth** area was settled by two separate but religiously similar culture groups. The Pilgrims who first settled the area around Massachusetts Bay in 1620 generally had little education or means but wanted to establish a colony based on their strict religious beliefs and to be free from persecution. Later they were to be joined by a different group who were quite well educated and from a higher social class. This hearth is unusual in that the second group, the Puritans of the Massachusetts Bay Colony, played a far more important role than the original group in shaping the region's economy and culture. The Puritans soon were both far more numerous and influential than the Pilgrims.

The **socio-economic culture** of these two groups was totally different from the tidewater colonies. The New Englanders settled in loosely organized villages, farm villages. They practiced a form of subsistence agriculture growing about as much food as the family could consume, with little surplus for sale. This practice has led to the people being very frugal and not allowing waste, a trait that has become indicative of the New England or Yankee culture. The people had an extremely high birth rate, and the rapid population growth forced settlement farther into the interior, gradually dispersing into individual homesteads. As settlement expanded, agricultural townships, with centrally located churches, a meetinghouse and all the amenities of a small town developed; while several large urban areas were established such as Boston, Salem and Providence. These places generated a strong demand for food, which encouraged nearby farmers to turn to commercial rather than subsistence farming.

The colony founded by William Penn, Pennsylvania, had liberal policies from the start on religious freedom and immigration, and it attracted a more varied European population than most other colonies and soon had a diverse cultural economy. Farmers specialized in growing grain, wheat and corn. This hearth region became the bread basket for the whole of the east coast. **The Midlands** came to symbolize the heterogeneous, yeoman-farming, free labor culture of early American ideals and practices that were to spread with the growth of population beyond the Appalachian Mountains.

Borrowing from Aboriginal Cultures

The process of cultural interchange between the aboriginal Indian population and the infiltration of the colonial newcomer has had a varying degree of influence on American culture. The original contact between the early colonists and the aboriginal culture groups, although important for the very early pioneers, failed to bring about a lasting aboriginal contribution to North American culture that would be distinctly insignificant.

It is estimated that at the time of early colorizations there were somewhere between 2 million and 4 million aboriginal Indians in North America. These people lived in a wide variety of environments, with hundreds of distinct societies and highly differentiated in language and culture. Some were hunter-gathers while east of the Mississippi River the Native Americans practiced a fairly intensive agriculture using simple tools, such as digging sticks and fire. In the Southwest, some tribes had developed sophisticated irrigation systems that led to larger and more dependable food supplies, and populations grew quickly to match the enhanced output. Wherever they lived, American Indians substantially modified their environments. There are recorded incidents of early explorers seeing large areas of intensive agriculture. Both the open eastern forests and the prairies of Indiana and Illinois, where forests had once stood, give strong evidence of native impacts on the environment by the use of fire.

Cultural integration came about with the development of the European cultural areas. The Natives developed a strong demand for many European goods, such as tools, blankets, weapons and decorations. In return, Europeans wanted some goods the natives could provide, particularly animal skins and furs. A vigorous trade developed well before extensive European settlement, an exchange with far-reaching consequences.

The early British cultures were being established in a backwater of Indian settlements. However, when the British came they found only remnants of much larger tribes that had been reduced by warfare and disease. They strongly believed that they could make much more effective use of the land and found little fault in taking it, for they believed their God would approve. A belief reinforced by the settlers' low opinion of the people they called heathens, who were ignorant of the Christian god and salvation. Only after the British had established their colonies did they come across Indians with a social order or farming practices approaching their own. By the time this came about it was too late to formulate a workable sociable apparatus to accommodate the two groups. No serious attempt was made to force a plural society. Neither was there an amalgam of bloodlines and cultural heritage as has occurred in Latin America, New Zealand or Hawaii. With the unfortunate conclusion that once the Republic was established there were only two approaches that were deemed to be effective; outright extermination or corral the survivors into reservations.

If we ask what has been the contribution of the American Indian to American culture, some might say very little. However, linguistically there are many instances of Native Indian words in the language. Geographically we see their influence in place names and early route ways. Twenty-six states have Indian names including Massachusetts, Connecticut and the Dakotas. River names such as the Penobscot, Kennebec, Merrimac and Connecticut in New England plus others such as the Susquehanna and Cuyahoga are derived from the Native Indians. Other Indian words that have been assimilated into the language describe indigenous species of tree, such as *hickory* and *pecan*; domesticated plants such as *maize, squash, sunflowers*; and animals such as *chipmunk, moose, terrapin and racoon*. Descriptive words like *totem, papoose, squaw, moccasin, tomahawk, igloo* and *kayak* have been passed down. We also see phrases and catchwords related to or derived from Indian culture, such *pow-pow, warpaint, play possum* and *go on the war path*. In addition, Indian lore and knowledge of nature was passed on in such forms as dyes, fibres and poisons. Although it might be true to say that had the European settlers found a completely empty continent contemporary American life would differ little from its actual modern pattern, we do have some linguistic evidence of intermingling of the two cultures.

The Local Evolution of American Culture

Once the various culture groups began to emerge and diffuse from their hearth regions they began to evolve into a single national culture by several processes. Acculturation took place by the integration of selective cultural traits from the old country. Some withered and died, others gained strength from their survival value such as; woodcraft, hunting and fishing.

Initially the isolation and distance separating different groups caused cultural isolation in different regions, however, once pioneers spread out across the country new social and economic patterns emerged. Others were governed by the larger national cultural system. The spatial juxtaposition of social and ethnic groups, which had been widely separated in the Old World, led to a spontaneous cultural interchange. Spatial reshaping of old world elements yielded interesting cultural consequences. If we consider a man as part of a complex bio-system then interaction is a two-way phenomenon. Man had an effect on the land but he was guided to a certain degree by the biota in which he existed. To this effect we should consider other influences on the development of the culture. Besides the British influence, the impact of the African slave and those peoples from other European countries should also be considered.

The African natives, nearly all slaves, were present in significant numbers from 1619. At the time of the Revolution they accounted for one-fifth of the total population of British North America. Under these conditions it would be unlikely that there was no cultural impact. It has been written that the African slave arrived in a state of cultural nakedness or amnesia, that he jettisoned his African heritage because he was incapable of practicing any of it in this new land. Despite the obvious forced acceptance of a non-African language and culture some remnants of Africanism remain. There are pockets where the African culture is strong, particularly where there was a demographic domination. For example, the Sea Islands off South Carolina and Georgia have a dialect—Geechic—which is as much West African as it is English. The same can be said of the Creole speech used in Louisiana.

While there are few recognizable impacts of the African culture in North America, there is relatively little African influence on the cultural landscape for instance. The African culture does predominate in American popular culture as dance, music, coiffeur and dress. Few words have been admitted into the language although today **Ebonics** has surfaced in the larger inner cities. African

Americans who have returned to Africa are actually aware of the cultural gulf between the two sides of the Atlantic.

Whereas the early domination of the American culture was by European groups such as the French, Spanish, Dutch, British and German; in later years other peoples have brought their culture to North America. After the formation of the Republic there was a huge influx of immigrants from other parts of Europe, as well as from the British Isles. Although subsequent immigrants from Ireland, Southern Europe, Italy and Greece in particular, Eastern Europe and Russia have contributed in varying degrees to the cultural mix, it is difficult to assign a particular innovation or trait to a particular group. As they reached North America these new immigrants experienced cultural shock followed by total acculturation, if not in the first generation certainly by the second and third. Most of the antecedent cultures, for example the Florida Spanish or the California Russians, were sunk without a trace, while the French and Spanish of the interior have made little contribution; they were simply pulverized under the Anglo-American advance. In most of the larger cities culture enclaves survive; New York has Little Italy, Cleveland—Slavonic village, and San Francisco—Chinatown. There are a few survivors. The Cajun folk of Louisiana speak a French patois because they are isolated and came to the region under suffrage and have adhered to their heritage more rigidly. Also an Hispanic or Chicano culture has survived in the Southwest of the United States. In fact, the Southwest is the only example of cultural pluralism in the United States.

One of the pivotal facts of North American cultural diversity is that from the establishment of the republic onward, later arrivals such as those from southern and eastern Europe and Scandinavia were swamped by the well-established Anglo-Teutonic colonization. Why did the Anglo-Teutonic cultures succeed while others, such as the Spanish, fail? The Spanish found the region physically and economically uninviting. They had not reached a stage in their own economic development where large numbers could afford to emigrate. On the other hand, the British found North America more attractive. They were economically and demographically ready for colonizing other parts of the world. They also had little competition by virtue of superior military, industrial and economic strength.

While the concentration of cultures has been on the east and south we must not forget the vital impact of cultures from the other side of the Pacific Rim. For instance, the Chinese played a significant role in the development of California in particular and together with the Japanese and Filipinos have had a significant influence on the culture of the western states.

In the later part of the 19th and early 20th centuries the new migrants had certain characteristics that can be attributed to their assimilation into the North American culture. These characteristics may not be the conventional attributes that we associate with the new migrants. The new incomers were not a representative sample of the people or culture back in the old country. They were largely young male adults—sometimes with children. They arrived mostly in single-family groups although there was some chain migration—other members of the family followed. Poverty was not the chief spoke driving people across the ocean, although there was a panic migration following the great potato famine in Ireland. There were more non-established church groups than established church groups. An example might be the Amish from Switzerland. Adherence to such "fringe" groups no doubt reflected and reinforced certain non-modal personality types or deviant political proclivities. Taking this into account, consider the following: Would the American Revolution have taken place had the new immigrants been a true cross-section of the original population?

A Continuing Interchange with Other Parts of the World

The umbilical cord between immigrant and homeland is seldom totally severed. Much returning migration takes place. More recently with a sharp acceleration in communications, commerce and ideas, and the circulation of visitors, the impact of the outer world on America has grown markedly. At the same time the effect of American culture on other groups has been magnified even more remarkably with the globalization of world economies and communications.

For four centuries, Anglo-America has witnessed revolutionary change not matched in man's existence on earth. Yet despite the mass movement of people and communications there still can be found, surviving in odd corners, some relics of early European cultures. For example, in the backwaters of Quebec, Canada, one can get a glimpse of life in 17th-century France; while in modern-day Pennsylvania 18th-century a Swiss-German lifestyle can be found amongst the older order Amish.

CULTURAL IDENTITY

"American culture, by its astonishing standardization of thought and uniformity of values in relation to size and population permits us to make generalizations which would be far more difficult to formulate for many other countries in western civilization" (R. M Huber, in Meredith, 1968).

A true ethnic group is also a true nation in the most primitive meaningful sense of the term. The *ethnic nation* may find political solace in the nation state though, if it is less fortunate, it will be submerged into a larger political entity and distributed over the territory of two or more such states. It has been suggested that the population of the United States is a genuine ethnic group recently fused from miscellaneous materials through a combination of spontaneous formulation and deliberate engineering by powerful socio-economic forces. Despite that there still exists a regionality to the cultural diversity within the country.

Spatial Patterns of Culture Groups in the United States

Many of the new immigrants moved out from this eastern seaboard and across the continent where large sections came to be dominated by one country or ethnic group. There are still large pockets of ethnic groups throughout the country. For example, in Minnesota and the Dakotas, there is a large Scandinavian population; Texas has large German enclaves, southwest Michigan has a Dutch influence, while in Pennsylvania and West Virginia's coal country, Slavonic groups predominate. Big cities too had their large ethnic settlements: Germans in Milwaukee and Cincinnati, Poles in Detroit and Buffalo, Irish in Boston and Cubans in Miami.

The availability of certain terrain, plant cover, soil and climate all reminiscent of the homeland or responsiveness to known skills may help create territorial zoning of ethnic groups. Once a viable ethnic nucleus takes a hold on a given location, chain migration may result. However, viewed at the microscopic level across the periods of immigration Europeans developed steadily outward. Initially large sections of the North American continent were dominated by one country or ethnic group. As movement continued westward this changed into county domination, in the plains states this is confined to towns, while in the far west group dominance seldom exists.

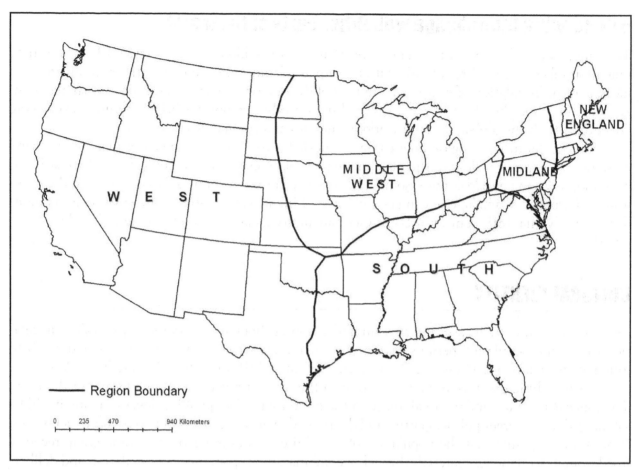

Figure 10.1   Cultural Regions. (University of Akron Produced)

The movement from the hearth areas took place over many decades and has resulted in a cultural mosaic that has some spatial significance. The dominance of the early cultural hearth regions, the South, the Midlands and New England still remains; to these have been added the Middle West and the West that emerge after the formation of the Republic. There is some argument for the Southwest, which is included in the West and part of the South, to be a separate cultural region based on the fact that it is a hearth for the many Hispanic culture groups that have developed and have influenced this area. Following are brief outlines of each of these cultural regions (see map, Figure 10.1).

The first of these cultural regions is **New England** with its Puritan culture and Yankee traditions; it has been a stopping-off point, the hearth, for cultural diffusion in the northern states. Although early settlement was from England, the region is dominated by people with Irish and German ancestry that today by virtue of speech, religious behavior and thought give the region its identity.

The Midland region is that area based around New York, New Jersey and Pennsylvania that emerged later in the colonial period by a mixture of Dutch, Northern Europeans and British. This mixture of European cultures created a distinctive cultural landscape based on a mixture of religious and economic influences. There are a variety of Christian sects including Quaker, Welsh Baptists, Catholics and Old Order Amish. The economic landscape varies from farmland, to coalfield, to factory. As the name suggests the Midland is intermediate in character between New England and the

South. Its residents are much less concerned with, or conscious of, its existence than is true for any of the other regions (Zelinsky, 1973).

The South in many people's minds is synonymous with the Confederacy. It is a development out of the Plantation economy of the early settlers and it is more a mixture of cultures than the other two prerevolutionary areas. We can identify various influences in the cultural dynamic of the region. There is the French influence in Louisiana, which was augmented in 1755 by the displaced Acadians from Nova Scotia who developed the Cajun culture. There were further French influxes from the West Indies before and after 1776. In addition, the region always had a Spanish influence from the earliest Spanish settlements at St. Augustine and more recent immigrants from Central America and the West Indies, notably Cuba. One further influence is those of African descent who have made their mark in many ways on the culture of this region. As a vernacular region, many of the inhabitants identify this region as Dixie, the area south of the Mason Dixon line. The region is noted for its vernacular architecture, language and cuisine, all of which stand out from the rest of the country.

Like the South, the **Middle West** enjoys a distinct identity. Most people living within or outside the region know of its existence. However, the determinations of its boundaries are a matter of conjecture. Where the region begins and ends usually depends on whom, and for what, the boundaries are being drawn. For instance, a person living on the west coast may see that part lying east of the Mississippi as the East, while a person from the eastern states may see the Midwest starting at the Mississippi. The epithet Midwest derives from the early days of interior colonization. To people moving west to build new settlements this was midway on their journey to lands west of the Mississippi. This region has in turn been the breadbasket of the nation and the industrial engine room. In the early days of settlement a mixture of German, Scandinavian and Slavic farmers settled what came to be known as the Corn Belt. Later, these people were to be joined by Southern Europeans and people of British origin from the East Coast, who arrived en masse to the industrial cities of Cleveland, Detroit and Chicago. These new immigrants brought new languages and religions to the region, including Lutheran to the northern states and Catholicism to the inner cities.

In the eyes of many **the West** is a land dominated by cattle ranching. Its history immortalized in Hollywood Western movies. It is true that much of the West has been built on cattle, but other areas have developed along other economic lines. Texas has its oil fields and California its gold-fields. Southern California, central California and the Mormon region of Utah and southern Idaho stand out as having a distinct cultural identity. Southern California until the late 19th century was remote, rural and largely inconsequential. The arrival of transcontinental railroad brought people from all over North America and transformed the area into one of avant-garde luxuriance. Here is a mixture of all races and religions found on the continent. Southern California has become synonymous with opulence, big cars and freeways. Central California is a land developed out of the gold rush of the mid-19th century, a frenzied time that like the southern part of the state attracted disparate groups from all over America and the rest of the world—especially the oriental Pacific Rim. From these groups has developed a cosmopolitan outlook that tends to look westward to countries on the other side of the Pacific, such as China and Japan. Economically the region has developed truck farming into an art form with many cities in the East dependent on the produce of this area. The Mormon region is emphatically based on a purely American religion—the Church of Jesus Christ of Latter-day Saints. Founded in upstate New York, it was carried west by its members seeking an isolated place to settle and follow their beliefs. According to Meinig, the Mormon region expresses a definite tribal group, with principles associated with an isolated culture. This complex group still appears to be non-western in spirit: The Mormons may be in the West but they are not entirely of it (Meinig, 1965).

The **Southwest** is a cultural region that stands out from the West because it is an established hearth area. This is the area around the Rio-Grande, southwest Texas, and stretches into Arizona.

Spanish speaking communities have been present here since the late 16th century; they developed a culture that has been augmented by continual immigration from Mexico and Central America. What is significant about this region is the resilience of the aboriginal cultures. Here the Navaho, Pueblo and lesser groups with their own distinct cultures continue in varies stage of assimilation. In some places the people have become strongly Americanized yet in others there are still undisturbed pre-Columbian cultural patterns. Given the special cultural qualities of this region and its general economic prosperity, it has one of the fastest growing populations in the country; it will continue to grow and attract new migrants from other parts of the United States.

PATTERNS OF INDIVIDUALISM

Although we may recognize several patterns of cultures on the landscape there is also a degree of individualism in American society. Individualism in human culture is a relatively new cultural trait; prior to its emergence, society was tradition bound. Individualism in the United States has expressed itself in several forms. One of these is the Protestant Ethic, a potent force in American economics and theology that manifests as a worldly salvation through constant industry and frugality. It is, in effect, based on the ethics of the New England cultural hearth that has insinuated itself across the greater part of the United States. Second is the Frontiersman, the resourceful isolated fighter against the wilderness, which translated into modern terms translates into the technological and scientific exploits overseas and in outer space. Some see this cultural expression as the new American. Finally, the Success Ethic is the simplified form of the Protestant Ethic, defined as fulfilling the greatest good to be the greatest possible individual success. Unfortunately, defining and measuring success has still to be worked out. Money or the acquisition of it is the numerical device used to keep score.

Despite a drive for individualism, Americans are very conforming. They turn to churches, political groups, lodges, clubs and other transient associations to fill the social gap left by modernity. The genuine certified eccentric who indulges his whims above a steady social base is almost unknown in America.

If individualism in American culture reveals itself in any geographical phenomenon it does so nowhere more strikingly than in the political landscape. In spatial terms the fragmentation of political authority is extraordinary and is never more evident than in the larger metropolitan areas where several hundred distinct governmental units may exist. In these larger metropolis' an individual may not only find territorial units such as city, village or township, but groups with special interest administrating a wide variety of social enterprises such as water supply, sewage disposal, pollution control, highways, harbours, schools, hospitals, airports, police, planning, zoning and parks. The list can go on and on. This crazy quilt of local authority reflects a deep-seated desire for individualism in America. What may sometimes be lost on many Americans is that this form of individualism costs money! In many instances the individual citizen likes to have control over these governmental bodies, exercised through the ballot box, but dislikes the extraordinary tax burden imposed by each specific group.

The compelling individualism and antagonism toward truly effective government is an integral strain on the cultural economic pattern. In areas west of the Appalachians the settlement patterns reflect this individualism. Isolated farmsteads abound. This isolation is very much an American model, as it does not reflect the clustered settlements that were the norm of European culture. These landscape patterns are the result of what Thomas Jefferson saw as his Agrarian Republic. As new lands passed to congressional control they were divided for sale into the system called Township and Range. It replaced the older metes and bounds system, which used natural features (e.g., rivers, rocks, trees) to demarcate property lines; that system dominates along the East Coast and in the Southeast. A township consisted of a square 6 miles (9.7 km) on a side that was divided into 36 "sections" of 1 square mile (640 acres or 259 hectares) each. A quarter of a section, or 160 acres (65 hectares), was considered the standard size for an American farm. Consequently, roads follow field boundaries usually in straight lines and a farmstead may be set in the middle of a quarter section, some way from its nearest neighbor. In this landscape there is no provision for towns and the roads. In fact the patterns make no sense; they do not cut down on distance but follow the rectangular pattern of the land subdivision (Peacefull, 1996).

To find further evidence of individualism in American culture, one should look no further than the areas of religion and education. There is no better example of individualism in the social fabric than with religion where so many different denominations exist that one family can be split into the different churches they attend. With education, individualism is represented in two ways. First, control of public schools is at the lowest local level, where there is little supervision by state government and almost none by federal government. Second, alongside public schools and colleges there is a proliferation of parochial and private establishments. A proliferation found nowhere else in the English-speaking world. These features of religion and education further attest to the nation's material wealth as well as to its individualism.

A Desire for Change

If one has to define a unique feature of the American Culture, it might be to consider America as an ongoing process—a culture that is forever in a state of flux. An innate restlessness and desire to change are underlying determinants of American culture. As a group, Americans are insensitive to or uninterested in any segment of time except the present or immediate future. Unlike other cultures, they do not dwell in the past, they demolish older artifacts and replace them with newer (better) ones. What evidence is there that New York City has existed for nearly 400 years? Only in recent years has there been a move to preserve the historical heritage. There is an innate desire for the novel and eagerness for built-in obsolescence from paper tissues to computer software and electronic gadgetry—some of which may not be justified on economic grounds.

Time is another factor, it is all-important—"time waits for no man, time is money!" Americans have translated linear distances into the temporal ones. Ask how far it is from Cleveland to Chicago and the answer invariably comes back about six hours. We build interstate highways and high-speed passenger jet planes to cut down travelling time. Yet we have yet to invest in a fast intercity rail network similar to those that are predominant in Europe and Japan.

America is forever on the move. More than 75 percent of all Americans over 18 change their residence more than once. In any given year about one in five Americans move house. There is a continuous cycle of movement: college, military service, marriage, employment, divorce and retirement; each will have a different spatial implication on an individual's life. Added to this is the daily movement to work. Only a tiny majority of people live within walking distance of their place of work. A majority travel 30 to 40 miles or more by car to their place of employment. Besides the daily commute, many travel across country on business, vacations and excursions to see family and friends. In fact for the average American, the state of physical rest seems to be the exception rather than the norm.

From the earliest settlement pattern we can observe the American devotion with mobility and change has been the catalyst that has developed the identity and cultural landscape. The variety of culture groups that have moved across the continent has brought about distinctive cultural landscapes in the form of the architecture and use of the land. In the next section we examine the impact of different culture groups on the North America landscape.

AMERICAN CULTURAL LANDSCAPE PATTERNS

Nothing tells us more of a culture than its buildings. A region's cultural history, ethnic origin and composition are contained in the built landscape. Certain structures clearly are associated with different ethnic groups and thus provide a means for identification and study of those people. The form of Ukrainian houses in Manitoba, the three-room arrangement of so-called Quaker-plan houses in the piedmont of North Carolina, Colonial houses in eastern Massachusetts, eastern Long Island and eastern New Jersey, show remarkable similarity, which reveals a common settlement connection. Many structures are identified by means of ethnic designations; "Dutch," "English" and "German Bank" are each clearly distinguishable barn types. Buildings are also a reflection of the resources available; for example, a line drawn between Princeton, New Jersey, and Wilmington, Delaware, would mark a general division between stone buildings to the north and west and brick buildings to the south and east. Such a differentiation reflects the geological resources of the two areas. In Lancaster County, Pennsylvania, the blue-gray limestone houses and barns produce a cultural landscape that is quite distinct from the patterned red-brick houses of Salem County, New Jersey. Less than 50 miles separate the two areas (Noble, 1984).

Besides being a reflection of the geology, buildings also reflect social status. Gowans (1964) notes in areas of Dutch settlement stone was used for dwellings occupied by people with lower incomes whereas brick structures housed upper-income groups. Besides status, fashion has a lot to do with the popularity of particular house styles. Almost every small town has a few 19th-century houses with elaborate "gingerbread work," a hallmark of the *Gothic Revival* style that became popular in the 19th century. As the North American economy shifted from an agricultural base to that of urban-oriented industry, so building styles changed from the slowly evolving traditional style to the rapidly evolving formal architectural styles that conformed to the fashion of the time; fated to last only a few decades, to be succeeded by some new fashion.

The Settlement Landscape

North America has many advantages and a few disadvantages for the study of the cultural landscape. The effective occupation of North America is a phenomenon of the last two centuries. Thus, in all sections of the continent, many of the original pioneering structures still stand. In addition, various culture groups from distinct areas of Europe tended to settle in separate regions in North America; creating clearly defined regional settlement groups consisting of houses, barns, churches, mills and smaller structures. Even in areas where settlement by various ethnic groups was not geographically separate, one group tended to succeed another so that periods of dominance of a particular group may usually be defined. In some areas the groups that arrived first occupied the land and succeeding groups went to villages and towns, effecting an almost immediate separation of cultures.

Because population was sparse and the land not completely occupied, it is relatively easy to trace the migration and assess the impact of ethnic groups by noting the distribution of structures or architectural features typically associated with each group. For example, migrants from the early colonial settlements around Massachusetts Bay made their way in the middle of the 17th century, first to Long Island and somewhat later to New Jersey. This migration is reflected in the Cape Cod cottages of New England, which are little altered on Long Island, and are still basically similar in New Jersey. These migratory patterns are further exemplified in the evolutionary sequence of New England-derived houses, which can be traced from their hearth, in New England, across central New York into northern Ohio and southern Michigan. Not all early original structures are museums or protected buildings. Careful research will identify private residences representative of early architectural types.

More houses than barns or other economic structures survive because the basic function of the house has remained unchanged. However, we can see the cultural origins of the early settlers more so in the agricultural structures than in the house forms. This is mainly due to the fact that architectural fashion of house design changes more rapidly than any other. To be sure, over time, barn structures have changed but at a slower pace than houses. Noble (1984) notes several cultural variations in North American barn types that have been transplanted from their ethnic European origins. The English barn is a simple rectangular structure with two or three bays. It is associated with early New England colonization and subsequent settlement of the interior. The Quebec long barn is an elongated version of the English barn found in the French parts of Canada. The German bank barn is a close cousin of structures found in the Germanic parts of Southern Europe. The latter has spawned many variants in North America, for instance, the Sweitzer barn, the Wisconsin Porch barn and the Pennsylvania H barn. As with houses, study of various barn types can identify the migratory patterns of the various early ethnic settlement groups.

There are two major disadvantages in studying the American cultural landscape. First, rapid population growth over the past 150 years has led to the destruction of early structures to be replaced with "modern" buildings. The replacement buildings are usually more efficient, but often less important architecturally, than the ones demolished. It is unfortunate that the most architecturally important buildings are found in the oldest parts of towns and thus most susceptible to modernization. Also, as the population grows and moves away from the city, earlier rural buildings are being replaced by modern suburban housing developments and related business plazas.

The second disadvantage and one of the characteristic features of the North American personality is the tendency, as noted earlier, to change and ignore whatever is not current or modern. In the process, many of the significant buildings that would help explain architectural evolution in the New World have been ignored and ultimately destroyed. Thus, a casual examination of eastern North America reveals a confusing array of material-cultural landscape features. Older structures are scattered amid newer ones and appear to be unrelated to their present surroundings.

Despite these disadvantages there is a regionality to North American settlement. In Quebec, Canada, the French settlers have left their mark with cottage-style houses that are a reflection on the climate. The Quebec cottage and the Montreal House are both one-and-a-half storey dwellings where the roof comes down to the first floor with a pitch between 45 and 60 degrees, significant in an area of heavy snow. Moving to New England we have already mentioned the Cape Cod house, but there are other styles that reflect the early English influence; the Salt Box, for instance, is a development for the growing affluent population who needed more space. This growing affluence led to New England's larger houses being based on the fashion of classic revival.

In the Hudson valley, the Dutch settlers left their mark with a different style of house. They also made use of brick as a building material. A typical Dutch colonial house would have a porch, usually the length of the first floor. These were the extension of the stoop—raised platform, usually with a bench either side, where guests were greeted and the weary traveler could shake the mud from his boots and coat.

The most widely distributed house type is the English I-house. These houses are elongated from the basic square of the New England homes. There is a central entrance way with separated living areas on either side instead of the one large living area. This house style was so popular that it spread far and wide throughout the United States—in Pennsylvania, across the Midwest as far as central Nebraska. Not only was it a widespread style, it continued to be fashionable up to the beginning of the 20th century, long after other styles had faded away. Many houses that appear to be of a distinct architectural style turn out to be I-houses.

As we move south into the tidewater colonies, house designs change as people's tastes were different. They also reflect the climate and environmental conditions in this part of the east coast. The coast is low-lying and marshy, leading to swampy conditions where malaria was endemic. Consequently, the house type reflects these conditions. These dwellings, known as the Charleston single house, were mostly one-storey, a single room in width to allow for cross ventilation, built of wood. A verandah known locally as a piazza ran the length of the house. Later variation on this style of house sees a second storey added, while in the city the house turned at right angles to the street. Although these house types are culturally distinctive they, along with those of the Spanish farther south, have not had a profound influence on later house architecture.

Moving along the gulf coast and into the Mississippi delta we come across the French influence on the landscape. The French had moved into the interior long before the British had established their colonies along the east coast; either moving down from the Great Lakes-St. Lawrence valley, or arriving at the mouth of the Mississippi from their Caribbean settlements. As with their occupation in the St. Lawrence valley, the French imposed the long-lot system of land divisions. This system attempts to divide the land so that each landowner has similar amounts of soil, landforms and vegetation conditions with access to a waterfront or other transportation route. Another French-speaking group that has influenced the architecture of the south is the Cajuns—the Acadians who were expelled from Nova Scotia by the British. These people brought a simple structure known by several different names as the Acadian house, the Creole house, or the grenier house. This is a distinctive structure of the delta region and bayou of Louisiana. The house usually sits on pillars, made of cyprus stumps originally, but more recently of cement block, that raise it several feet off the ground. The reason for elevating the building is not clear. Some speculate it was for cooling allowing air to circulate, while others suggest it offers flood protection. Another feature of the house is the porch or galerie thought to have been originally the sleeping area for the young men in the family.

Two further porched or galleried houses in the French colonial style are found in the Mississippi valley. One is simply a French-Canadian house with a porch and the other is the Louisiana plantation house. Like most houses in the lower Mississippi Valley, these were of wooden construction due to a lack of building stone in the area. One of the most distinctive house types of Louisiana, especially in the New Orleans area and other parts of the south, is the shotgun house. This design is derived from French cultural influence in the Caribbean and brought to the delta area by freed slaves. The house is remarkable for its narrow façade and great depth. The name is said to refer to the alignment of the doorways so that a shotgun fired at the entrance would exit out the back door. Noble suggests it is more likely to be a corruption of the West African word "to-gun" meaning a place of assembly (Noble, 1984).

In the interior, many varied dwellings appear on the landscape, many of which are variations on those found farther east. One dwelling of note is the sod house of the interior plains. In this part of the continent, building materials such as wood were scarce so the early pioneers use sod. The buildings were of two types: the basic house with four sod walls and a sod roof or the sod dugout, dug into the side of a ravine or valley. The entrance to the dugout and partial roof were usually sod. Thousands of the structures were built by the early land claimants who, once the claim was staked, had to have some shelter. Only in areas where there was an abundance of trees such as the Pacific Northwest were the more familiar log cabins built.

In the southwest at least two Indian cultures have persisted: the Navajo and the Pueblo. The Navajo constructed a circular building called a hogan, either from stone or logs. More recent forms have been polygonal in shape and built of sawn timbers. Modern hogans are constructed with manufactured materials such plywood with tarpaper roofing. The Pueblo Indians are named for their form of settlement. Pueblo is Spanish for settlement town or village. The buildings in which these people live is known as an adobe, which is also the name for the sun-dried clay bricks used in the building process. The typical adobe consists of two or more large buildings separated by a courtyard. They may contain many rooms and be several stories high. Besides the Indian dwellings there is also a Spanish-Mexican influence in the region. The Spanish adobe is usually a single-storey building with rooms surrounding a courtyard the whole being termed a casa.

CULTURAL DIVERSITY IN THE UNITED STATES—THE DIVERSITY OF LANGUAGE

One of the significant aspects of the cultural patterns in the United States is the diversity of languages. Many would consider that English is the official language; it is the language of government, business and mass communication, yet there is no official language in the country. If we consider language within the United States, we find that it is truly a polyglot nation. This lack of an official language has remarkable consequences and at times a strain on the official system. For example, recently in Lowell, Massachusetts, public school courses were offered in Spanish, Khmer, Lao, Portuguese and Vietnamese, and all messages from schools to parents were translated into five languages. The multilingual, polyglot, New York City education system has bilingual programs in Spanish, Chinese, Haitian Creole, Russian, Korean, Vietnamese, French, Greek, Arabic and Bengali. In most states, it is possible to get a high school equivalency diploma without knowing English because tests are offered in French and Spanish. In at least 39 states, driving tests are available in foreign languages; California provides 30 varieties, New York 23 and Michigan 20, including Arabic and Finnish. And, as required by the 1965 federal Voting Rights Act, multilingual ballots are provided in some 300 electoral jurisdictions in 30 states.

These and innumerable other evidences of governmentally sanctioned linguistic diversity may come as a surprise to the many Americans who assume that English is the official language of the United States. It isn't. Nowhere does the Constitution provide for an official language, and no federal law specifies one. The country was built by a great diversity of cultural and linguistic immigrants, who nonetheless snared an eagerness to enter mainstream American life. At the start of the 21st century, a reported 18 percent of all American residents spoke a language other than English in the home. In California public schools, one out of three students uses a non-English tongue within the family. The 21,000 students in Fairfax County, Virginia, schools speak more than 140 different languages, a linguistic diversity duplicated in many major city school systems.

Nationwide, bilingual teaching began as an offshoot of the civil rights movement in the 1960s, was encouraged by a Supreme Court opinion authored by Justice William O. Douglas, and has been actively promoted by the U.S. Department of Education under the Bilingual Education Act of 1974 as an obligation of local school boards. Its purpose has been to teach subject matter to minority-language children in the language in which they think while introducing them to English, with the hope of achieving English proficiency in two or three years. Disappointment with the results led to a successful 1998 California anti-bilingual education initiative, Proposition 227, to abolish the program. Similar rejection elsewhere—Arizona in 2000 and Massachusetts in 2003, for example—has followed California's lead.

Opponents of the implications of governmentally encouraged multilingual education, bilingual ballots and ethnic separatism argue that a common language is the unifying glue of the United States and all countries. Without that glue, they fear the process of "Americanization" and acculturation— the adoption by immigrants of the values, attitudes, behavior, and speech of the receiving society— will be undermined. Convinced that early immersion and quick proficiency *in* English are the only sure ways for minority newcomers to gain necessary access to jobs, higher education, and full integration into the economic and social life of the country, proponents of "English only" use in public education, voting and state and local governmental agencies successfully passed Official English laws and constitutional amendments in 27 states from the late 1980s to 2002.

Ethnic groups, particularly Hispanics, who are the largest of the linguistic groups affected, charged that the amendments were evidence of blatant Anglocentric racism, discriminatory and repressive in all regards. Some educators argued persuasively that all evidence proved that, although immigrant children eventually acquire English proficiency in any event, they do so with less harm to their self-esteem and subject matter acquisition when initially taught in their own language. Businesspeople with strong minority labor and customer ties and political leaders, often themselves members of ethnic communities or with sizeable minority constituencies, argued against "discriminatory" language restrictions.

Historians noted that it had all been tried unsuccessfully before. The anti-Chinese Workingmen's Party in 1870s California led the fight for English-only laws in that state. The influx of immigrants from central and southeastern Europe at the turn of the 20th century led Congress to make oral English a requirement for naturalization, and anti-German sentiment during and after World War II led some states to ban any use of German. The Supreme Court struck down those laws in 1923, ruling that the protection of the Constitution extends to all, to those who speak other languages as well as to those born with English on their tongue. Following suit, state courts have also voided some of the recent state language amendments. In ruling its state's English-only law unconstitutional, Arizona's Supreme Court in 1998 noted it "chills First Amendment rights."

To counter those judicial restraints and the possibility of an eventual multilingual, multicultural United States in which English and, likely, Spanish would have coequal status and recognition, U.S. English—an organization dedicated to the belief that "English is, and ever must remain, the only official language of the people of the United States"—actively supports the proposed U.S. constitutional amendment first introduced in Congress by former senator S. I. Hayakawa in 1981 and resubmitted by him and others in subsequent years. The proposed amendment would simply establish English as the official national language but would impose no duty on people to learn English and would not infringe upon any right to use other languages. Whether or not these modern attempts to designate an official U.S. language eventually succeed, they represent a divisive subject of public debate affecting all sectors of American society.

Consider the following:

1. Do multiple languages and ethnic separatism represent a threat to America's cultural unity that can be avoided only by viewing English as a necessary unifying force?
2. Could making English the official language divide its citizens and damage its legacy of tolerance and diversity?
3. Would immigrant children learn English faster if bilingual classes were reduced and immersion in English were more complete?
4. Would a slower pace of English acquisition be acceptable if subject-matter comprehension and cultural self-esteem are enhanced?
5. Could official English laws inflame prejudice against immigrants or provide all newcomers with a common standard of admission to the country's political and cultural mainstream?

DISTINCT CULTURE GROUPS (1): THE AMISH

Of all the many different cultures within the United States, one, if not the most distinctive group, is the Amish. With their distinctive dress and mode of transport they stand apart from mainstream America. They base their culture and way of life on strict biblical teachings that control their daily activities, their clothing (plain with no fasteners), their use of hand tools rather than mechanically driven ones and their strict adherence to abstinence and observing the Sabbath. Although in some ways, as we shall see, the youth are allowed more freedoms than perhaps their American contemporaries.

It is estimated that each year over 10 million tourists visit the major enclaves of the Amish in Pennsylvania, Ohio and Indiana. Numerous tourist services have sprung up, among them restaurants, craft and antique shops, and the usual "authentic" Amish farms, complete with buggy rides. The local Amish appear to have accepted all this as inevitable. For some it actually presents a financial boon because tourists are all important purchasers of their homemade quilts, furniture and baked and canned goods. For some of the casual visitors 'Amishlands' are like visiting a foreign country without having to go abroad. The people have a different lifestyle, they speak a different language. It is an idyllic land. The tourists share the opinion that Amish life nurtures qualities that they do not have in their own life: simplicity, a slow-paced life style and "a oneness" with nature.

The Amish way of life is based on the tenet that they are charged with the task of being "in the world but not of the world"; maintaining that separation requires a constant battle with the surrounding culture, and in that endless conflict it is the depth, strength and completeness of their beliefs that sustain them (Schachtman, 2006). They set themselves apart by dressing in a special way, using horses as draft animals both in the fields and for transportation, rejecting the use of electricity, worshiping in their homes rather than in church buildings, ending their children's formal schooling after the eighth grade and engaging in many other nonmainstream behaviors because these activities are mandated by and in accordance with their religious beliefs.

The origin of the Amish church dates from the late 17th century Anabaptist or Swiss Brethren movements of Europe. The Anabaptists split from the main body of the Christian church over the right to baptize adults only and to separate church from state. The followers based their faith on seven tenets. First, that baptism must be achieved by choice of the adult. The second and third deal with discipline necessary to maintain "purity" of the church; the brothers and sisters of the sect hold one another accountable for their behavior which includes obeying the rules. Today this is known as *ordnung*.

The church backs these rules up by a system of punishments each more severe than the previous. The final admonition is to publicly disgrace the culprit and exclude them from the congregation—this is known as the bann or shunning—it is the harshest punishment that a sect can mete out. It is harsh because it virtually excludes the shunned from having contact with other members of the sect including members of the immediate family.

Today's Amish are the religious descendants of Jacob Amann, who, sometime during 1693–1697, broke with the Mennonites to form the Amish church. In fact, it was the rigidity of the bann that caused Joseph Amman to permanently split his followers from those of Menno Simons— the Mennonites. These two sects originated in the German Swiss area of Europe. Persistent persecution dispersed the Amish from Switzerland into the Rhineland in Germany, Alsace-Lorain in France and the Netherlands. Eventually forced from these areas, they migrated to America, initially settling in southeastern Pennsylvania among other German-speaking peoples. Both groups have managed to maintain many of the distinctive cultural traits, such as language and architecture, of their home region. However, the Mennonites enjoy greater integration into mainstream America,

Figure 10.2   Amish Buggy. (© 2011, Leonard Peacefull)

which sets them apart from their Amish cousins. While the Amish adhere to strict religious tenets, the Mennonites drive cars, use electricity and allow their children to continue in school, but continue to dress in the traditional way.

The Amish are socially organized in church districts, which they also call settlements. Between 25 and 30 farms may belong to a single district. Church services are held every two weeks and alternate between the various farms of the district. Although very community oriented, the Amish are not a communal people. In fact, they are very good capitalists who save primarily to purchase land for their children. Amish families are quite large, and the youngest child usually inherits the homestead, a traditional practice known as *ultimogeniture*. The Amish suffer from two main problems: insufficient land and outside cultural pressures. These problems often force families to move, leading to the recurring establishment of new settlements (Wilhelm and Noble, 1996). It is reported that a new Amish settlement is established somewhere in the United States and Canada every five weeks.

The Amish Cultural Landscape

The Amish have placed a certain characteristic on the Cultural Landscape of North America. Economically the Amish and their Mennonite cousins concentrate their resources predominantly in farming, although furniture making and construction also have some significance. Agricultural activity among the Amish centers on a five-year rotation of corn, wheat, oats, and grass. Farms are relatively small (average 80–100 acres). Any larger area could not be easily handled by their horse teams.

Figure 10.3   Amish House. (© 2011, Leonard Peacefull)

Amish farmsteads nearly always include two houses: the granddaddy or "grossvader" house and the main house. The former may be attached to the main house and usually shelters the retired elder farmer and his wife. There are also many outbuildings, including a large barn. This serves two purposes; first the storage of grain and hay, together with stabling and stall area for the horses and cattle. Second, is to provide space for church services, which the farmer is expected to do periodically. If the congregation is large they will not all fit into the farmhouse, so the barn is used. Besides being large, Amish residences are usually distinguished by the lack of wires connecting the house to poles along the road. For religious reasons they decline to use machinery unless sanctioned by the church as necessary to carry out their trade. This is particularly so in dairying areas where in order to sell milk for consumption they have to have a refrigeration unit.

At the beginning of the 20th century, there were as few as 5,000 Old Order Amish in the United States and Canada. However, from 1950 onward, along with much of the population their numbers increased rapidly, by 1980 this number had grown to 100,000 and in the early years of the 21st century this number has doubled to 200,000 individuals almost half of whom are under the age of 18. This doubling of the population is due to the high birth rate, low infant mortality rate—the average Amish family has seven children—and a very high retention rate, i.e., once baptized few leave the church. Today there are over 300 church districts in 28 states and the province of Ontario in Canada. Most are concentrated in Ohio, Pennsylvania and Indiana. An article in the weekly newspaper, the *Budget,* suggested that rate of growth of the Amish is the same as that of the Jews in Egypt: Joseph arrived with 12 men and 72 women and children; 12 generations later Moses led 600,000 out of Egypt.

Rumspringa

One surprisingly unique aspects of Amish culture is **rumspringa.** This is a Pennsylvania Dutch term, usually translated as "running around." A more exact translation would be "Running around outside the bounds." The rumspringa period begins when an Amish youth turns 16; at that age, since the youth has not yet been baptized, he or she is not subject to the church's rules about permitted and forbidden behaviors. During rumspringa Amish youth, for the first time in their lives, go on their own in the outside world. On weekends during what they call "going-away" they party, drink alcohol, smoke cigarettes and do most of the things in which mainstream American teenagers indulge. They also experience the opposite sex. On returning to the family farm after a weekend of partying most of the young Amish will not tell their parents precisely where they have been the previous 48 hours, or with whom they spent their time. While the parents may well ask questions, the children feel under no obligation to answer them.

Rumspringa ends when the young Amish decide to enter the church and become baptized. It is of interest to note that after this experience 90 percent of the young adults decide on baptism. Many older Amish, although not fully approving of Rumspringa, accept that this period in the life of youth allows them to see life outside the sect. It also allows the sect to keep a hold on many of its youth who might otherwise leave and reject the way of life. (See Schachtman, 2006, for more information.)

DISTINCT CULTURE GROUPS (2): THE CAJUNS

The Cajuns are a distinctive culture group of North America synonymous with southern Louisiana. To appreciate the Cajun culture, one must first understand the unique culture of their predecessors, the Acadians. These people can trace their history back to the early 17th-century Maritime Provinces of Canada when a group of French settlers arrived in the area that is known today as Nova Scotia. These settlers were mostly seeking relief from economic and religious oppression in western France. Upon arrival, they integrated with the indigenous Micmac people, learning how to survive along with the customs and ways of life of the native Indians. One notable aspect of the native culture which these French settlers soon adopted was that all people were considered equal and power was awarded through merit, not inheritance. Being largely isolated from their French homeland, the settlers developed an identity of their own. They called their new land "L'Acadie" after the Micmac word for "land of plenty." They became known as "Les Acadiens," or "Acadians." Though they still had ties to France, they no longer considered themselves French. Their culture was an indisputable product of the new world, combining many cultural influences to create their own way of life.

The Acadian Cultural Landscape

The Acadians chose a relatively isolated existence mainly to preserve their own cultural identity but also they did not want to be seen as having allegiance to whoever was in control of the area at the time—be it France or England. To affirm this, the Acadians persevered at a subsistence level.

The Acadians were an extremely frugal people; whenever a farmer needed a new rake, harrow or dray, it was more economical for him to make it from materials at hand than to buy one. Agriculture was also closely linked to clothes and fabrics; the women grew flax for linen and raised sheep for wool—basic materials for clothes making. Acadian women had their own spinning wheels, looms and other equipment made by their menfolk. Any other special equipment that was needed was most likely also built by the men. They were mostly illiterate as reading did not serve any immediate purpose in their agrarian society.

Religion, specifically Catholicism, was important to the Acadians and formed the basis for a large family-centered social unit. Acadians believed that success came from hard work and discipline, not from political or social connections. Over the course of the 17th century, the people developed a way of living that centered on God, the family and the land. This pattern served them well until 1755.

Le Grand Derangement

During the 17th and 18th centuries, control of Nova Scotia changed hands many times from France to England, yet the Acadians maintained their identity in spite of this turmoil. Never formally swearing allegiance to either nation, they were under suspicion from both sides. In 1755 the English, who thought the Acadians allies of France, decided to rid the area of a potential threat. The English imprisoned the Acadians; confiscated their possessions and forced them onto ships bound for English Colonies in North America and the Caribbean. Approximately 7,000 were exiled, others, about 1,000, fled to safety in the backwoods of what is now New Brunswick (Leblanc, 1970). The Acadians who were distributed through the American colonies were not well received. Many of the colonies, including Virginia, Georgia and North Carolina, either aided or completely funded attempts by the Acadians to return to their homeland.

Settlement in Louisiana

Among the Acadians who were spread throughout the Eastern Seaboard and the Caribbean, there remained a constant dream of finding a new homeland where they could reunite and nurture their culture once again. The area they chose to settle was the lands west of the Mississippi delta, later to become Louisiana, that became a focal point for many of those exiled from Maritime Canada. There are conflicting schools of thought as to why the Acadians chose this area. One argument is that they were driven by a motivation to perpetuate their cultural identity. In 1764 the Spanish, who controlled the area were eager to fill it with anyone other than English settlers, welcomed the Acadians. The Acadians then began arriving in this new land from the eastern colonies, the Caribbean Islands and Europe. The last major group arrived on seven ships from France, hired at the expense of the Spanish government in 1785. Eventually, about 4,000 Acadians reached Louisiana.

Once in Louisiana, the Acadians again found themselves in a new and very different environment from Nova Scotia. The leaders of the groups chose for settlement the relatively isolated Attakapas district because of its suitability for agriculture and husbandry. This region lies to the west of the Mississippi River in an area covered by both swamp and prairie lands and is isolated from the Mississippi by the Atchafalaya swamp. Here they began again to establish a homeland by farming and raising cattle; they were able to preserve their way of life away from other cultural influences. The geography, language and occupational opportunities tended to reinforce the social isolation and protect the Acadians from cultural changes. A French administrator of the Spanish territory at the time commented that if Acadians continued to arrive, then Louisiana would become the new Acadia.

Diffusion of the Cajun People

For the next few generations following the arrival of the Acadians in Louisiana, the culture slowly metamorphosed into its present Cajun state. There are few landmark events that indicate a definite turning point, but several changes can be identified that mark the change. Toward the end of the 1700s and into the early 1800s, new Americans began to move into the Louisiana territory. These people tended to be more educated and socially aggressive than the Acadians who had chosen to remain socially isolated and largely uneducated. As a result, these new Americans began to acquire prime lands and essentially force the Acadians farther from the ever expanding centers of society.

The environment of southern Louisiana was very different from maritime Canada and the people had to adapt. Different types of crops and animals, clothing and housing styles were all indicative of changes in the Acadian lifestyle. Four different environments were inhabited by the Acadians that influenced the development of the Cajun culture.

The levee lands along the Mississippi and the Bayous were usually the most sought after and prosperous lands. These lands provided Acadian settlers access to waterways that were used for transportation and irrigation, but as the land was seen as the most productive, the levees were the first to draw the attention of the new American immigrants.

The prairie lands of southwest Louisiana were some of the original lands to be settled by the Acadians. These lands allowed the Acadians to maintain their culture unchanged from their Canadian roots. However, the environment was very different and they had to adapt both the types of crops grown and animals raised. Over time, two distinct cultural zones emerged. One zone was the corn and cotton section and the other, the rice and cattle section (Conrad, 1978). The corn and cotton region was the most fertile region, while the rice and cattle section consisted of less desirable, clay-like soil.

The swamp lands offered displaced Acadian agriculturalists the opportunity to develop new ways to earn a living. The Acadians became swamp dwellers and fishermen. They abandoned their traditional homes for life on a houseboat and became fishermen and purveyors of swamp goods such as Spanish moss, frogs, crayfish, turtles and crabs. The marsh environment was the last environment settled because its lands were mostly underwater and not particularly suited for agriculture or husbandry. These lands were essentially a last resort for settlers.

As the Acadians began to spread across the southwest portion of Louisiana, and the Anglo-Americans moved in behind them, the identity of the "Cajun" began to emerge. The term "Cajun" was derived from the English pronunciation of Acadian and was generally a pejorative term seeking to define the Acadians as lazy, uneducated, stubborn and self-centered citizens relative to the upwardly mobile, business savvy Anglos. In reality, the Cajuns were, by choice, a socially isolated group that as a whole chose to center their lives on God, their families and the land.

In reality, the Cajuns were and are a welcoming people. They were a very cooperative society with a very strong system of reciprocity. Much of the labor-intensive work of the farms was done collectively by the male members of a family or parish. House moving, building structures and bringing in crops are all examples of these activities. One well-known Cajun custom was the boucherie—the rural butchery. Because of the heat and humidity of southern Louisiana, fresh meat would only be preserved for a few days at most, so parish families would take turns "sponsoring" the boucherie and men from the surrounding area would participate in the slaughter and in return, each participating family would get a portion of the freshly butchered meat—enough to last until the next boucherie. These gatherings not only served a practical purpose, but served a social purpose as well. Meetings such as these were a gathering place where news of the parish could be shared and local residents would find out which of their neighbors were ill, who had died, who had been born and who had fallen on hard times and might need additional assistance.

Other cultural activities that helped solidify group cohesiveness included cooking, bals de maison (folk dances), veillées (evening gatherings), coups de main (cooperative work) and weddings. The cooking and preparation of meals also served as a social occasion where guests were expected to assist in the cooking of outdoor meals such as shrimp, crab or crawfish boils in the summer and gumbos in the winter. The "coup de main" also aided in strengthening the community spirit of the Cajuns. If a Cajun was too ill to work, his neighbors would pitch in to help. All members of the group were expected to participate when aid was required and as such, a sense of trust developed. All of these cultural events played a role in strengthening the Cajun culture in southern Louisiana.

Cajun Language, Food and Music

When considering the language of the Cajun people, we can identify three distinct varieties of French. First, there is the formal Standard Louisiana French. Second, there is Acadian or Cajun French, the most widespread variety, even with its various dialects. It is understood by most French-speaking Louisianans. The difference between Cajun and Standard French is a result of the influence of the Spanish, English, Caribbean and African languages also spoken in Louisiana. Finally, there is Creole French or "Gumbo French" which is spoken by relatively few people.

Cajun food is an example of the thrift and resourcefulness of the people. Much like the examples from other areas of their lives, the Cajuns used their Acadian roots and added and intermingled foods, spices and techniques from other cultures to create their own unique style and flavor of food. The roots of Cajun cooking are based on the premise that nothing should go to waste. Cajun dishes feature sauces, which are cooked and simmered for an extended period of time. These techniques were employed in order to improve the tenderness of the lesser-quality meat being used and to allow flavors to blend, helping mask any questionable flavors in the meat. These meats were then paired with one of the staples of the Cajun diet, beans. Ham and/or smoked sausage with rice or beans are standard Cajun fare and are representative of dishes served in parts of southern households. One signature ingredient in Cajun cooking is the crawfish, a crustacean that has only recently become popular outside of southern Louisiana. When people outside of Louisiana think of Cajun food their mind turns to Gumbo. Gumbo has its origins in Acadian thrift and includes whatever ingredients are at hand, excluding beef or ham unless they are in the form of fresh or smoked sausage. The main ingredient of any gumbo is okra, a vegetable originally imported from Africa that flourishes in the fields of Louisiana. The gumbo is served over a bed of rice.

Cajun music, like their food and language is a blend of differing styles. Cajun music, originally adapted French folk songs, reflects the influences of German, Scots, Irish, Spanish, Native American, Afro-Caribbean and Anglo-American music. The additional influences are also reflected in the instruments. The instrumentation of Cajun music in the early 20th century included the fiddle, the guitar and the accordion. These instruments represented the influence of the French (fiddle), the Spanish (guitar) and the Germans (accordion). Toward the end of the 20th century there was a revival in the interest of Cajun music, with bands and musicians participating in folk music festivals across the country.

The Cajun Culture Today

Change came to Cajuns during the middle part of the 20th century. Soldiers returning after two world wars brought an awareness of modern conveniences not found in the levees and bayous. The demand for electricity, telephones, television and cars increased. The Cajuns began to understand the need for formal education as a means to better employment and a better, albeit more Americanized, lifestyle. People began to desire the trappings of modern life, but at the same time, they were not willing to cast aside their unique culture. This new awareness began a cultural revival.

There was a movement to put the French language, no longer being looked upon as a "lesser" language, back into the school systems. Cajun food became popular nationally and internationally while Cajun music enjoyed renewed attention as major record labels began to record and promote Cajun artists.

After almost 400 years of almost constant pressure to conform, today the Cajun people are acquiring their own cultural independence. They have proven their identity is too strong to be eradicated. External challenges enhanced by the developments of the present century have taken their toll on many subcultures. In Louisiana it appears that these challenges have largely been met; with many young people now taking pride in their heritage where there once would have been shame. The story of the Cajuns is one that is an inspiration to other cultures as a story of survival and adaptation.

KEY TERMS

Yankee Culture	Cultural Interchange	Cultural Identity
Ethnic Nation	Tidewater Colonies	coureurs de bois
Spatial Patterns of Culture Groups	New England hearth	The Midlands
Middle West	Socio-economic culture	Patterns of Individualism
Settlement Landscape	Ebonics	Rumspringa

DISCUSSION QUESTIONS

1. In what ways did the arrival of the early European settlers (French, Spanish, Dutch, and British) influence the evolution of American culture?
2. What effect did the process of cultural interchange have between the early settlers and the aboriginal population? Discuss with respect to culture traits.
3. Taking the early cultural evolution of the continent into consideration and had the new immigrants of the 1700s been a true cross section of the population is it possible that the revolution of 1776 would not have taken place?
4. Do the cultural regions as expressed by Zelinsky and paraphrased here still exist?
5. Besides the Amish and the Cajuns, are there other distinct culture groups in North America? What makes them distinct?
6. How does the Amish Rumspringa period compare to American teenage years?

References

This is a list of the works sited in the text.

American-Israeli Cooperative Enterprise, The. 2011. www.Jewishvirtuallibrary.org.

ARDA 2011, http://www.thearda.com/QuickLists/QuickList_125.asp

Azad India Foundation. 2010. *Gender Inequality*. http://www.azadindia.org/social-issues/GenderInequality.html.

Bednarz, Sarah Witham, Schug, Mark, Miyares, Ines M., and White, Charles S. 2005. *Social Studies: World Cultures and Geography*. Houghton Mifflin, Boston, MA.

Billikopf, Gregorio. June 1, 2009. "Cultural Differences? Or, Are We Really That Different?" June 1, 2009. http://www.cnr.berkeley.edu/ucce50/ag-labor/7article/article01.htm.

Burchell, R. W., Lowenstein, G., Dolphin, W. R., Galley, C. C., Downs, A., Seskin, S., Still, K. G., and Moore, T. 2002. Costs of Sprawl—2000. *Transit Cooperative Research Program (TCRP) Report 74*, Washington, DC: Transportation Research Board.

Conrad, Glenn R. 1978. *The Cajuns, Essays on Their History and Culture*. Lafayette: Center for Louisiana Studies, University of Southwestern Louisiana.

Crews, Angela. April 2000. "Understanding Cultural Differences." http://www.professionalroofing.net/archives/past/apr00/international.asp.

Deetz, James. 1977. "Material Culture and Archeology—What's the Difference?" pp. 9–12 in Ferguson, Leiand (ed.). *Historical Archeology and the Importance of Material Things*. Lansing, MI: Society for Historical Archeology, Special Publication, no. 2.

Earle, Francis M. 1943. "The Japanese House," *Education*, 63:277–81.

Esfandiari, Golnaz. 2004. *Self-Immolation of Women on the Rise in Western Provinces*. http://www.dhushara.com/book/sakina/stoningetc/immol.htm.

Gale, Fay. 1993. A View of the World through the Eyes of a Cultural Geographer. In Rogers, A., Viles, and Goudie (eds.). *The Student's Companion to Geography*. Oxford: Blackwell.

Golany, Gideon S. 1992. *Chinese Earth-Sheltered Dwellings*. Honolulu: University of Hawaii Press.

Goodman, Ellen. 1983. Take Back Your Quiche Monsieur, *Boston Globe*, March 8th 1983.

Gould, Peter. 1993. *The Slow Plague: A Geography of the Aids Pandemic*. Cambridge: Blackwell.

Gowans, Alan. 1964. *Images of American Living,* Philadelphia: J.B. Lippincott Co.

Grigg, D. B. 1974. *The Agricultural Systems of the World—An Evolutionary Approach.* Cambridge: Cambridge University Press.

Hägerstrand, T. 1953. *Innovation Diffusion as a Spatial Process.* Translated by A. Pred. Chicago: University of Chicago Press, 1967.

Haggett, Peter. 2001. *Geography a Global Synthesis.* Harlow: Pearson Education.

Henderson, V. 2002. Urban Primacy, External Costs, and Quality of Life. *Resource and Energy Economics,* 24: 95–106.

Hewitt, Duncan. 2009. They're Not Going to Take It. *Newsweek 8/01.*

Hill, R. C., and Kim, J. W. 2000. Global Cities and Developmental States: New York, Tokyo and Seoul. *Urban Studies,* 37: 2167–2195.

Hitchcock, Susan Tyler, and Esposito, John L. 2004. *The Geography of Religion, Where Gods Lives, Where Pilgrims Walk.* Washington, DC: National Geographic Society.

Humphrey, Caroline. 1974. "Inside a Mongolian Tent." *New Society,* 30: 630:273–75.

Jackson, John B. 1984. *Discovering the Vernacular Landscape.* New Haven: Yale University Press.

Jacob, John S. 1995. Ancient Maya Wetland Agricultural Fields in Cobweb Swamp. Belize: Construction, Chronology, and Function, *Journal of Field Archaeology,* 22(2), pp. 175–190.

Johnson, Paul. 1987. A *History of the Jews.* New York, NY: Harper and Row.

Johnston, R. J. 1993. *The Challenge of Geography.* Oxford: Blackwell.

Johnstone, B. 2000. The Individual Voice in Language. *Annual Review of Anthropology,* 29, pp. 405–424. Palo Alto, CA: Annual Reviews.

Kinzer, Stephen. 2008. *Crescent and Star: Turkey between Two Worlds,* 2nd ed. New York: Farrar, Straus.

Kushner, Harold, 1993. *To Life: A Celebration of Jewish Being and Thinking,* Boston MA, Little Brown

Leblanc, Robert G. 1970. "The Acadian Migrations." *Canadian Geographic Journal,* 81.1:10–19.

Lewis, Jonathan. 2006, prod. "Women of the Country." *China From the Inside.* KQED/Granada Television.

Lewis, M. Paul (ed.). 2009. *Ethnologue: Languages of the World,* 16th ed. Dallas, TX: SIL International. http://www.ethnologue.com/.

Mango, Andrew. 2004. *The Turks Today.* Woodstock, New York: Overlook Press.

Marsh, George Perkins. 1864. *Man and Nature: Physical Geography as Modified by Human Action.* New York, NY: Charles Scribner & Co.

McCrum, Robert, Cran, William, and MacNeil, Robert. 1986. *The Story of English.* London: Faber and Faber.

Mead, R. W. 1954. Ridge and Furrow in Buckinghamshire, *The Geographical Journal,* 120(1), pp. 34–42.

Meinig, Donald. 1965. "The Mormon Culture Region: Strategies and Patterns in the Geography of the American West 1847–1964." *Annals of the Association of American Geographers,* 55(2).

Meredith, Robert (ed.), 1968. *American Studies: Essays on Theory and Method*. Columbus, OH: Charles Merrill Publishing Company.

Morrill, R. L. 1970. The Shape of Diffusion in Space and Time. *Economic Geography*, 46, pp. 259–265.

Noble, Allen G. 1984. *Wood, Brick, and Stone: The North American Settlement Landscape*. 2 Vols. Boston: The University of Massachusetts Press.

Noble, Allen G. 2007. *Traditional Buildings*. London: I. B. Tauris.

Oliver, Paul. 1987. *Dwellings: The House Around the World*. Austin: The University of Texas Press.

Peacefull, Leonard. 1996. *A Geography of Ohio*. Kent, Ohio: Kent State University Press.

Peacefull, Leonard. 2011. Seeking the Light in Modern Turkish Relationship Issues. *The Arab World Geographer*, 14(1).

Queen Rania of Jordon. 2007. *My Message on Cross Cultural Understanding*. Speech delivered at The Women's Conference, California.

Quraishi, Ahmad. 2011. *Self-immolations Increase in Heart*. http://www.rawa.org/temp/runews/2011/03/30/self-immolations-increase-in-herat.

Religious Tolerance 2011 http://www.religioustolerance.org/worldrel.htm

Rich, Tracey R. 2010. www.jewfaq.org.

Rubin, Rehav. 1991. Settlement and Agriculture on an Ancient Desert Frontier. *Geographical Review*, (2), pp. 197–205.

Sauer, Carl O. 1925. "The Morphology of Landscape." In *Land and Life: A Selection from the Writings of Carl Ortwin Sauer*, ed., by J. Leighly. Berkeley: University of California Press, 1963, pp. 315–350.

Sauer, Carl O. 1969. *Seeds, Spades, Hearths, and Herds*. Cambridge, MA: MIT Press.

Sauer Carl O. 1971. The Agency of Man on the Earth, in Thomas et al. *Man's Role in Changing the Face of the Earth*. Chicago, University of Chicago Press.

Schachtman, T. 2006. *Rumspringa—To Be or Not To Be Amish*. New York: North Point Press.

Stark, Rodney, and Bainbridge, William Sims. 1987. *A Theory of Religion*. New York: Peter Lang.

Stump, Roger W. 2008. *The Geography of Religion, Faith, Place and Space*. Plymouth, UK: Roman and Littlefield.

Ullman, E. L., and Boyce, R. R. (eds.). 1980. *Geography as Spatial Interaction*. Seattle: University of Washington Press.

UNESCO 2008. *Operational Guidelines for the Implementation of the World Heritage Convention*, Annex 3, Paris, World Heritage Centre.

Union for Reform Judaism. 2008. www.Reformjudaism.org.

United States Environmental Protection Agency (USEPA). 1997. *Health and Environmental Effects of Ground-Level Ozone*. http://www.epa.gov/region7/air/quality/o3health.htm.

United States Environmental Protection Agency (USEPA). 2008. *Municipal Solid Waste Generation, Recycling, and Disposal in the United States: Facts and Figures for 2008*. http://www.epa.gov/osw/nonhaz/municipal/pubs/msw2008rpt.pdf.

Vearey, J., Palmary, I., Thomas, L., Nunez, L., and Drimie, S. 2010. Urban Health in Johannesburg: The Importance of Place in Understanding Intra-Urban Inequalities in a Context of Migration and HIV. *Health & Place*, 16: 694–702.

Warrick, Catherine. 2005. "The Vanishing Victim: Criminal Law and Gender in Jordan." *Law and Society Review*, 39.2 315–348.

Wilhelm, Hubert, and Noble, Allen. 1996. "Ohio's Settlement Landscape" in L. Peacefull, *A Geography of Ohio*. Kent, OH: Kent State University Press.

Zelinsky, Wilbur. 1973. *Cultural Geography of the United States*. Englewood Cliffs, NJ: Prentice-Hall.